Behind the Screen:
The American Museum of the Moving Image

Written by David Draigh

Introduction by Rochelle Slovin

Guide to who does what in motion pictures and television

American Museum of the Moving Image

Abbeville Press · Publishers · New York

THE PENNSYLVANIA STATE UNIVERSITY
COMMONWEALTH CAMPUS LIBRARIES
DELAWARE COUNTY

D1317091

Library of Congress Cataloging-in-Publication Data
Draigh, David, 1963-
    Behind the screen:... who does what in motion pictures
and television.
    "Published to accompany the permanent exhibition
entitled 'Behind the screen' at the American Museum of
the Moving Image, Astoria, New York, September 10,
1988."
    Includes index.
    1. Motion picture industry—United States—Job
description. 2. Television broadcasting—United States—
Job description. I. American Museum of the Moving
Image. II. Title.
PN1994.D68   1988   384'.8'023   88-6183
ISBN 0-89659-955-8 (Abbeville Press : spiral bound)

Copyright ©1988 by American Museum of the Moving
Image. All rights reserved under International and Pan-
American Copyright Conventions. No part of this book
may be reproduced or utilized in any form or by any
means, electronic or mechanical, including photocopy-
ing, recording, or by any information storage and retrieval
system, without permission in writing from the publisher.
Inquiries should be addressed to Abbeville Press, Inc., 488
Madison Avenue, New York, N. Y. 10022. Printed and
bound in the United States of America. First edition

Designer: Stephanie Tevonian, Works
Editor: Georgette Hasiotis
Picture Researcher: Shari Segel
Production Manager: Dana Cole

The publication of <u>Behind the Screen</u> heralds the opening of a new national institution: the American Museum of the Moving Image, the first museum in the world dedicated to exploring the full network of forces that create – and are created by – the moving image media. The subject matter of the Museum is <u>all</u> moving images, from classic films to television commercials, from documentaries to the avant-garde.

This volume and the Museum's permanent exhibition of the same title are designed to answer the question, "What are all those jobs listed when the credits roll?" By presenting a remarkable range of film and television artifacts, from a full-scale movie set to interactive exhibits and specially commissioned artists' installations, the exhibit encourages the public to take a step behind the screen and meet the many talented collaborators – actors, directors, editors, scenic artists, publicists, theater owners, network executives – responsible for producing, promoting, and exhibiting films and television programs. The exhibit makes it clear that motion pictures and television shows do not spring fully realized from a single mind. They are created through a complex set of interactions among a host of talented people, whose skills, experience, and ideas are significant only in relation to the contributions of others.

One of the Museum's prime goals is to provide the public with new ways of seeing the moving image media. The remarkable synthesis of art, craft, technique, technology, culture, and history that we plan to reveal has not been explored in other institutions. Since any critical assessment must first consider the collaborative nature of film and television productions, <u>Behind the Screen</u> is a most appropriate beginning.

The job descriptions that follow use both "he" and "he or she" in stating duties and functions, since there is no general-use, nonsexist pronoun. But the choice of pronouns largely reflects the reality of the proportion of men and women working in the fields described. It is true that many jobs in film and television have traditionally been dominated by men; however, most of the early screenwriters were women, and in the 1920s many talented women moved from jobs as negative cutters, script keepers, and typists to become film editors. There are still many women film editors, some at the top of their field. Today, women are working in all areas of the motion picture and television industries, in settings as diverse as the major studios, networks, and local television stations, and in jobs that range from director to gaffer.

The job titles in this book are those most commonly used in film and television credits for each area of responsibility, but there are variations since some jobs simply have more than one name and others have changed or evolved over time. Where significant, these variations are pointed out either in the heading or in the text itself.

The moving image industries are perhaps the most talked about in the world. However, as relatively new media they are not that easily understood. The preparation, production, and dissemination of films and television shows involve prolonged processes of considerable complexity. Behind the Screen attempts to get past the film industry's traditional glamour to demystify the inner workings of what may be the most mystifying of all art forms.

—Rochelle Slovin, Director, American Museum of the Moving Image

Pre-production
production
post-production

Most jobs in this book are said to fall into one or more of three distinct phases of production. Though the boundaries between these phases are somewhat flexible, the following terms have long been an indispensable part of the industry's vocabulary:

Pre-production is the planning phase, and encompasses a range of activity including budgeting, scheduling, casting, set and costume design, location scouting, set construction, and special effects design.

Production is the "lights, camera, action" of Hollywood legend, when the cameras roll and the ranks of cast and crew swell sometimes into the hundreds and beyond. Production can take as little as one day in television, as much as several months or more in film.

Post-production is when activity moves from the set to editing rooms, scoring stages, recording studios, and laboratories, where the project is shaped and molded into its final form.

Unions + Guilds

Although there are many exceptions in independent, documentary, and similarly small-scale production, most mainstream film and television production is handled by union personnel. Some unions serve specialized segments of the workforce, such as the Directors, Writers, or Actors guilds. Others, such as IATSE or NABET, cover an extremely wide range of crafts and technical skills, from makeup artist to director of photography to grip to scenic artist.

The jobs described in this book are followed by their union affiliations. Workers in jobs represented by more than one union do not belong to all of them. Union membership, sometimes a matter of choice, is often dictated to some degree by geography or place of employment. Most union activity is found in major production centers such as New York, Los Angeles, and Chicago. The full titles of the unions, in alphabetical order, are:

AFM	American Federation of Musicians
AFTRA	American Federation of Television and Radio Artists
DGA	Directors Guild of America
IATSE	International Alliance of Theatrical Stage Employees
IBEW	International Brotherhood of Electrical Workers
IBT	International Brotherhood of Teamsters
NABET	National Association of Broadcast Employees and Technicians
SAG	Screen Actors Guild
SEG	Screen Extras Guild
USA	United Scenic Artists (a local union of the International Brotherhood of Painters and Allied Trades)
WGA	Writers Guild of America

Actors are so much a part of our culture – in the movies, on television, in magazines and newspapers, and in all kinds of advertisements – that we sometimes take their presence for granted. But this has not always been the case. The very first actors had far less exposure and didn't receive so much attention. In the early years of cinema, film acting was considered a poor cousin to stage acting, and actors often chose to remain anonymous and uncredited so as not to spoil their chances in the legitimate theater. This arrangement suited producers as well: The anonymous actor didn't command a star salary and films were promoted more by the name of the studio than by the actors they featured.

Public interest in the faces on screen grew rapidly, however, and soon popular actors found themselves in a position to bargain. When Mary Pickford negotiated a contract in 1916 giving her $10,000 a week, the star was born.

Yet the people with their names in lights are only a small portion of the acting profession's working corps. For every starring or supporting role, there are thousands of bit parts, walk-ons, and fleeting appearances requiring the actor's unique skills.

Some actors prepare for a career by attend-

1

ing schools and academies; others may get a break without this formal training. However, most actors hone and perfect their craft throughout their career, for example, by continued private instruction and in workshops.

1 The stars come out in celebration of MGM's twentieth anniversary, 1944. Studio chief Louis B. Mayer sits at front center.
2 Students play a scene for their classmates and Director Sam Wood at the Paramount Acting School, Astoria Studio, 1925.
3 Studying makeup technique at the Paramount Acting School, Astoria Studio, 1925. In the early days actors applied their own.

2

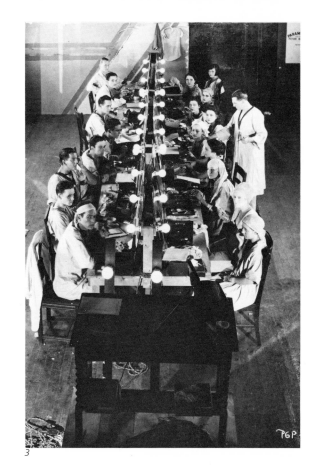

3

7

Now, as in the past, screen actors often get their start in the theater. Today they are likely to divide their time between film, television, and the stage. The different requirements of stage versus screen acting require that an actor be flexible. While the stage actor performs in narrative sequence within the space of a few hours, the screen actor performs in scenes that are probably out of narrative or chronological sequence and possibly shot over a period of months — yet this fragmented performance must be delivered so that it will make sense when edited into sequence.

Once a part is landed, the actor may supplement his training with research about the character he will play. On the set, an actor's closest working relationship is with the director, but by the time a movie is complete, the actor's image is truly a product of the many crew members with whom he has worked so closely throughout production — the director of photography, sound mixer, hairstylists, makeup artists, and costume designers — all of whom help create the character we see. **(AFTRA, SAG)**

Fencing class at the Paramount Acting School, Astoria Studio, 1925.

The ADR (for automatic dialogue replacement) editor is responsible for the post-production completion or alteration of a film production's dialogue tracks. Production dialogue tracks might need doctoring for any number of reasons: excessive location noise, scenes shot without sound, or the director may simply decide to try to get a better reading from the actors. When the film editor has completed the final cut, he or the sound editor will contact the ADR editor, and together they determine the portions of dialogue – from a word here and there to whole passages and scenes – that need work.

In a specially equipped studio the film segments (usually no more than ten seconds) are projected on a screen while the actor, wearing a headset that plays the existing dialogue track, speaks his lines to match the lip movement on screen. The ADR equipment allows this process to be easily repeated, a valuable feature, since success is rare on the first try.

The ADR editor and his assistants handle all preparations for the ADR sessions, including breaking down the picture and sound tracks into the segments that will be used, preparing cues for the actors, and planning and scheduling the sessions.

By the time ADR is under way, principal photography is long finished and the actors may be thousands of miles away. It is usually most economical for the ADR editor to travel to the actors rather than the other way around. The director is usually present at ADR sessions to coach the actors and decide which reading will be used. (IATSE, NABET)

Advisor/Consultant

Films and television programs explore subjects ranging from war to medicine to baseball, but their makers are rarely generals, doctors, or first basemen. In addition to research in books and periodicals, the advice of experts in other fields may be required to ensure the accuracy of what goes on screen. These advisors may work with the director, the actors, the production designer, the costume designer – anyone whose job requires a more exact knowledge of the subject or setting of the production. Their contribution may be limited to advice given during pre-production or extend into the shooting phase.

Agent

Agents were an established part of the film industry even during the silent era. In the teens and twenties they began to represent actors in contract negotiations and writers wishing to sell scenarios to the studios. Through the thirties and forties they achieved a greater presence and a broader base of clients and, with the decline of the studio system in the fifties, their power greatly increased. As the studios dropped their contract talent rosters, agents' client rosters swelled, and producers often depended on agents or agencies for whole "packages" of a property, director, writer, and stars. In such cases, the agent or agency would be acting in the capacity of producer and, in addition to the percentage fees collected from their clients, would receive a portion of the film's profits.

This system remains much the same today, and agents have broadened their client base even further to include production designers, directors of photography, editors, costume designers, choreographers, etc., as well as actors, directors, and writers, for whom they also sometimes assume the role of career manager.

Animal "actors" — from Rin Tin Tin to Lassie, Flipper, and Elsa the Lioness — sometimes gain greater fame than their human co-stars. Whether winged or four-legged, famous or not, screen animals are rarely far from their trainers, who specially prepare them both to perform their required actions and to work calmly and reliably in an environment full of lights, cameras, people, and other potentially scary distractions. The trainer also has full responsibility for the animals while on the set or location, coaching them, caring for them, and ensuring that working conditions are safe.

On a production requiring an animal "star" or many animals, the trainer is contacted early, allowing time to study the script and shooting schedule and to prepare the animals — sometimes with the help of as many as fifteen or twenty assistants. Yet it is not unusual for a trainer to receive a call requesting anything from an elephant to an emu for a scene the following day. (IBT, NABET)

1 *Her Jungle Love*, 1938.
2 Between takes on the set of *The Arabian Shrieks*, Astoria Studio, 1932.

2

1

The subject of animation deserves a book unto itself, and many have been written. The classic form of American animation, as exemplified by the Disney classics and used for most Saturday morning cartoons, is cel animation. Other forms are object, clay, and puppet animation.

In cel animation, the animator draws figures and backgrounds onto separate, transparent cels, which are stacked on top of one another when photographed. The number of layers of cels depends on the number of moving parts in any given scene. For example, if a background is stationary but a character runs across it, the animator will draw a succession of cels representing the different positions of the character, but need draw only one cel of the background.

The puppet animator makes minute, painstaking adjustments of movable puppets, each of which are recorded by the camera frame by frame. Standard camera speed is twenty-four frames per second, so one second of screen time requires twenty-four adjustments. Any object, from rocks to pencils, can be animated using this same procedure. The art of the animator has also been applied to live action, frequently in special effects movies. For example, the familiar laser beam that destroys an enemy alien or spaceship is probably the work of the animator, who draws the beam to match the previously filmed live action.

Though an animated feature film represents the contributions of dozens of animators working under a supervising director, animation is a form of filmmaking that may be practiced by independent animators who perform every aspect of the process, from concept to drawing and photography. (IATSE)

Walt Disney *and staff gather in celebration of his 1932 special Oscar for Mickey's creation.*

Art Director

To avoid any confusion at the outset, the title "art director" until recently designated the person now referred to as the "production designer." Likewise, the art director used to be called the "set designer." The change in terminology has evolved over the past twenty years, though in many television productions the title "art director" is still used to designate the head of the art department.

The art director is responsible for organizing the art department into an efficient unit for accomplishing the design work – color and surface textures, plans and elevations, working drawings, etc. – as envisioned by the production designer. With the designer and the production manager, he or she assembles the necessary crew – set decorators, draftsmen, scenic artists, and perhaps a storyboard artist. Throughout the design process the art director is the production manager's right hand, and must be capable of conveying the designer's ideas to other members of the art department at every step along the way.

The art director also works with the property master on the selection of props and acts as a liaison with the construction department.

Most art directors have architectural training of some kind and attain their position after working as a draftsman, scenic artist or assistant art director. The next step up for the art director would be the position of production designer. (IATSE, NABET, USA)

Assistant Art Director

The assistant art director helps the art director carry out the design concepts envisioned by the production designer. His primary responsibility is building scale models of the sets. Scale models, a significant, time-tested aspect of the design phase, allow ideas to be evaluated and refined three-dimensionally by the designer. At various points the models are presented to the director and producer for approval.

Preliminary scale models are likely to be simple, hastily fashioned paper models; later models may be highly detailed, painted, and decorated prototypes . The assistant art director may contribute to the selection of color, surface texture, and architectural detail, all of which can be reviewed in the more elaborate models. (IATSE, NABET, USA)

On the set, the assistant director's authority is exceeded only by that of the director and the director of photography. The assistant director must know, in detail, the director's wishes for every aspect of the production, and he or she communicates much of the director's vision to the film crew.

Early in pre-production the assistant director and the production manager plan every element of the shooting, from preparing production breakdowns (stunts, sets, cast, etc.) to location conferences and the hiring of crews and equipment. The assistant director determines the number of extras needed and usually hires an extra casting director to find them.

His or her most important pre-production task is the formulation of the shooting schedule. The director, producer, production manager, and various department heads all contribute, but the assistant director is ultimately responsible for devising a schedule that will maximize the production's resources of time and money. The factors to be considered include actor availability, time and budget limitations, the logistics of location shooting, and the weather.

During shooting, the assistant director distributes the call sheets, which are extracted

from the master shooting schedule. The call sheets provide each day's vital information — the scenes to be shot, the cast and crew required, where to be and when to be there. Working alongside the director, the assistant director acts as a stage manager. He or she works to keep everything running smoothly by communicating continually with just about everyone on the crew and set.

The assistant director has creative control over the direction of the background action — extras, animals, vehicles — while the director concentrates on the principal actors. It is the assistant director who calls for quiet on the set and for the cameras to roll, though the director always calls for action. (DGA)

Harold Bucquet *chalks in the call board on location for* The Pagan, *1928. The remote location—the Tuamotu Archipelago—accounts for the trilingual sign.*

....

Associate Director

On taped television productions, the associate director is the director's right hand through all phases of production. As the show begins to take shape, he keeps detailed notes of the director's staging decisions and camera choices, annotating the script with instructions. He organizes this information in written form for the technical director and uses it to prepare a "shot card" listing only those shots for which each camera operator is responsible. The associate director may also lead meetings with camera operators to explain the shots.

The associate director remains in the control room with the director and communicates with other key personnel by headset during rehearsals and taping. The most frequent voice on the headset, the associate director gives cues to camera operators and relays instructions from the director to the technical director, stage manager, and other production team members.

The associate director often plays an active role in the post-production editing of a television production. He assists the director in the selection of shots and, as the director's representative, may work with the videotape editor to put the show together. (DGA)

Atmospheric Effects Specialist

Film and television productions depict the full range of Mother Nature's moods, from gentle to violent, but the weather conditions called for in the script rarely prove amenable to filming or taping. When photographed, real rain tends to make the image look scratched, and it is rather difficult to shoot a picture in the middle of a hurricane, not to mention the fact that shooting under actual weather conditions means waiting for them to happen, a luxury few budgets can afford. When a script calls for rain, wind, snow, fog, or hurricanes, the producer usually engages an atmospheric effects specialist.

The atmospheric specialist generally starts work in pre-production by meeting with the director or producer to determine the nature of the effects and work out a budget. When the effects are extensive, he is usually responsible to the director; for small scale effects he may report to the production manager. The atmospheric specialist often requires a large staff of technicians to assist with the operation of several standard pieces of equipment – wind and fog machines, smoke pots – and on occasion equipment custom-designed to suit the needs of a particular production. The specialist may create rain by suspending pipes dotted with holes over a soundstage set, through the use of hand-held hoses, or even by submerging water jets in a giant water tank disguised as a lagoon. He may also rig up huge water tanks designed to release thousands of gallons of water to simulate a flood.

The atmospheric specialist must take pains to ensure that no one is overcome by effects such as smoke or fog; for potentially more hazardous effects he may be required to arrange for paramedics or the presence of local authorities. (IATSE, NABET)

1 Snow in the desert for Ice Station Zebra, 1968.
2 Cliff Robertson gets the dousing in a scene from Man on a Swing, 1974.

2

1

Responsible for all sound recorded during a television production, the audio operator gets the script well in advance to determine his needs in personnel and equipment. Throughout readings and rehearsals, he follows the director's blocking and staging decisions closely, both to know where the sources of voice and sound will be and in order to alert the director if a particular setup presents a recording problem.

The audio operator uses boom, hand-held, or small radio microphones attached to a performer's clothing, and has a full staff of assistants to operate them. During rehearsals and taping, the audio operator is stationed in the control room or in an audio room, communicating by headset with his assistants and other key personnel while he monitors and balances the sound picked up from the set. (IATSE, IBEW, NABET)

3 Monitoring sound on the CBS telecast of " Mr. Roberts," c.1952.

3

Best Boy

As chief assistant to the gaffer, the best boy must be a highly skilled electrician. Ideally he knows the skills of the gaffer so that he can work ahead of the film crew, preparing the lighting equipment as required for the next set while the gaffer remains behind to supervise activity on the set in use. (IATSE, NABET)

. . . .

Boom Operator

The boom operator manipulates the microphone boom, a mobile extension arm that allows him to position the microphone above or beside the performers. He works under the direction of the sound mixer and must follow the performers through complicated shots while making certain that the boom and microphone do not come into camera range or cast a shadow. His skill is an important factor in the quality of the sound recorded. (IATSE, NABET)

1 *In the NBC television studio, c.1955.*

2 *Breaker Morant, 1981.*

3 *Elephant Man, 1980. Director David Lynch kneels behind the bed.*

4 *Doctor Detroit, 1983.*

1

2

3

4

Cable Puller

Cable pullers, working in both film and television, are responsible for setting up and handling power, microphone, and video cables during filming or taping. Television cameras are linked to control-room monitors and tape recording machines by heavy cables carrying the video signals; if a cable becomes tangled or kinked, the signal may be interrupted and the camera put out of commission. The cable puller (usually one for each camera) prevents this by following the moving camera and keeping the cables free of interference. In film production, the cable puller may be required to handle both camera and sound cables (which run to the power source) during moving shots. (IATSE, IBEW, NABET)

1 In New York for North by Northwest, 1959.

2 On location in Arizona for The Garden of Allah, 1936.

3 Suitably attired on desert location for The Scalp-hunters, 1968.

2

3

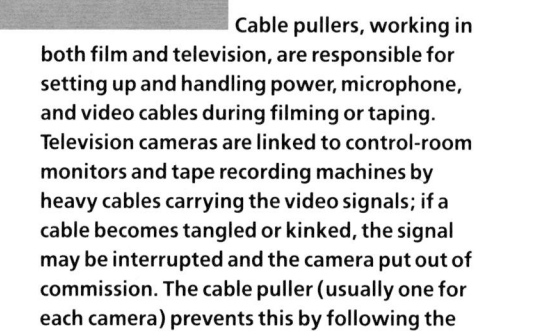

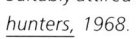

1

The camera operator is responsible for all facets of the operation of the camera, especially the smooth execution of the camera movements determined by the director and the director of photography. The operator participates in all camera rehearsals, during which the director of photography instructs him on the requirements of each shot. The rehearsals, sometimes conducted with the actors, sometimes with stand-ins, give the crew the opportunity to practice the take in one or more dry runs. During shooting the operator lines up the shot for composition, checks that the correct lens is in position and properly focused, starts the camera and checks that the film is running at the proper speed, makes sure the other camera crew members have done their jobs, and performs such movements as pans and tilts. When the camera must "travel", it is mounted on a dolly and pushed along tracks by the dolly grips.

The operator must have a great deal of technical expertise because he oversees the maintenance and repair of a wide range of sophisticated equipment. Several special-use cameras, for example, Steadicams, cameras for helicopter mounts, motion-control, crane-

4 *The troop of camera operators needed for* The Ten Commandments, *1923. Head cameraman Bert Glennon is on horseback.*

4

21

mounted, and remote-control cameras are often operated by specialists under the supervision of the director of photography. The camera operator has usually been a first or second assistant camera operator before reaching this position, which is itself often preliminary to a a position as director of photography. **(IATSE, NABET)**

1 *Coming in for a close-up of Dustin Hoffman and John Malkovich for the CBS production of "Death of a Salesman," 1985.*

2 **Alvin Wyckoff** *turns an early hand-held camera on W.C. Fields for It's the Old Army Game, 1926. Director Eddie Sutherland stands at center.*

1

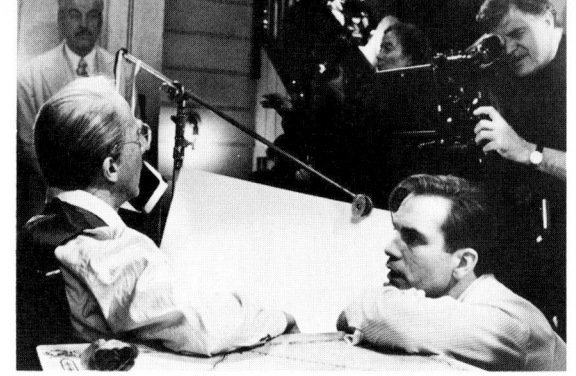

2

Although similar to the work of an operator on a film production, the television camera operator's job falls within the framework of a different technology and a significantly different production system. Taped television shows almost always involve multiple cameras, each with its own operator, and require a careful plan, almost like choreography, to determine which portions of the action each camera will record. The plan corresponds to the director's staging, and the associate director prepares individual shot cards listing the shots required of each operator. On an unpredictable production such as a live broadcast of a sporting event, the director makes camera decisions on the spot and communicates his instructions to the camera operators through the associate or technical director. Since the director's crew is often out of sight in the control room, the camera, sound, and stage crews communicate with the camera operator by means of a headset or intercom system.

The camera operator must line up shots, assist the video operator with camera set-up, mechanical checks, and maintenance; and is also responsible for controlling camera movement, lenses and focusing. On news pro-

3 A CBS mobile unit covers the Easter crowd on Fifth Avenue, c. 1950.

3

grams camera operators go into the field with reporters, where they use portable equipment and may also be responsible for sound recording.

Camera operators often have experience as production assistants or engineering technicians. A successful camera operator often goes on to become a lighting director, audio or video operator, or stage manager. (IATSE, IBEW, NABET)

1 *The CBS broadcast of the 1981 Western Open golf tournament.*
2 *Henry Fonda in the CBS telecast of "Mr. Roberts," c.1952.*

2

1

Carpenters work under the construction manager or foreman to build sets and flats for film and television from the working drawings prepared by the production designer or art director. They begin work during pre-production, when set designs have been approved and shooting schedules finalized. Carpenters may continue to work through at least part of the shooting phase, since shooting can take place on one set while others are under construction.

Carpenters also work on location, where environments are altered by the addition of false facades or whatever else may be required to make them more suitable. In television, carpenters work on extremely tight schedules, building sets in units or sections that can be easily stored and reassembled. (IATSE, NABET)

3 Putting together a sphinx for The Ten Commandments, 1923.

4 Building a set on location in Italy for Romola, 1925. The set, covering seventeen acres, was modeled after a Renaissance palace.

The casting director, once a studio executive overseeing the casting of several productions, is now more likely to work out of a casting agency, signing on to films or television programs on a project-by-project basis. In either case, the casting director is responsible to the producer for tailoring a cast to fit the budget, and to the director for assembling a cast that is right for the script.

Casting begins with the leads since they are the anchors of the ensemble, and following them in appropriate order are supporting players, bit parts, and extras. In consultation with the director, the casting director develops a list of possible actors for the lead roles. At this time, he conducts interviews or auditions with actors, sometimes taping them for the director to see. The process is repeated for the supporting players and bit parts, with the added dimension that they must complement the actors selected for the leads.

Casting a television program usually falls into two distinct phases. The producer hires a casting agency to perform original casting – assembling the group of regulars for the series. Once established, a staff casting director or casting agent performs the job of assembling performers for guest appearances and smaller parts on individual episodes.

The casting director is the producer's repre-sentative in contract negotiations and is, if necessary, the person who must go to the studio, producer, or network asking for more money, then ask the actors' agents to settle for less, until an accord is reached. He must be thoroughly familiar with pay scales and the salary histories of his actors.

1 **Fred Datig** *surrounded by screen hopefuls in the Paramount casting office, Astoria Studio, c. 1929.*

2 **Arthur Jacobson** *directs Florence Farley in a screen test at the Astoria Studio, c. 1935.*

Caterer

The production manager or location manager hires the caterer to keep film and television crews well fed on remote locations and, for the sake of convenience, in the studio. Catering companies maintain the tables, benches, tents, fleets of trucks, and other equipment that make them virtual restaurants on wheels in nearly any location, season, or climate.

3 *Production Manager Joseph Cook hands a ceremonial plate to Nelson Eddy in this publicity still from* Rose Marie, *1956.*

4 *Lunchtime for the thousands on location near Bakersfield, California for* Cimarron, *1931.*

3

4

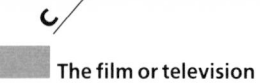

The film or television choreographer's job varies significantly from that of the stage choreographer because of the presence of the camera — camera height, angle, and movement must all be considered when planning dance routines. A choreographer is usually hired to plan extensive dance sequences in a production and works closely with the director throughout both planning and shooting. The choreographer is involved with the casting of dance parts, including principal performers, stand-ins, and chorus parts. The choreographer may also work with the production designer concerning plans for sets and locations where dance numbers will be performed. He or she usually requires a considerable amount of off-set rehearsal time with the dancers before shooting begins, and is present when dance numbers are performed.

1 **Busby Berkeley** rehearses Judy Garland for a scene in Babes on Broadway, 1941.

2 **Busby Berkeley** explains a dance routine in this elaborately posed publicity still from Golddiggers in Paris, 1938.

3 **Robert Alton** takes Lena Horne through a dance routine for Till the Clouds Roll By, 1946.

4 **Bob Fosse** works with Shirley MacLaine on the set of Sweet Charity, 1969.

5 **Jerome Robbins** makes a point on the set of West Side Story, 1961.

1

2

3

4

5

Of the many lab technicians who handle and process incoming negative, the color timer has the most control over the final result. The quality of his or her work critical to producing a positive print with the color values and tones desired by the filmmakers.

The producers contract with an outside laboratory to handle all film processing. The color timer analyzes the output of each day's shooting take by take, frame by frame, taking into consideration camera reports or the instructions given by the director of photography. He or she must achieve the intended effect as well as ensure a visual consistency throughout the film, especially between individual shots taken under varying light conditions. The color timer achieves this through the use of filters and by adjusting light intensity levels within the printing machine, which is hooked up to a computer. He or she creates a program for the computer that instructs the printer to make adjustments at the correct intervals in the film. The results of this process are the dailies, from which the filmmakers select takes for the final version of the film.

When editing is complete, the color timer repeats this procedure for the whole film, this time producing the answer print, so called because it is sent to the filmmakers for approval or disapproval. The color timer makes adjustments until the print meets the requirements of both the filmmakers and the distributor. Once approved, the lab prints the film in quantity for distribution and release. (IATSE, NABET)

1 **Annemarie Rzepka** *in the color timing department at Movielab, 1975.*

1

Films have an almost musical structure before even a note of the score is recorded. The edited images have their own pace and rhythm, and the music added to the images can serve a number of functions – enhance emotion, create suspense or surprise, reinforce or contradict what is happening on screen.

The addition of music is one of the last steps in a film or television production. A

2 **Maurice Jarre** *at work on his Oscar-winning score for* <u>Lawrence of Arabia</u>, *1962.*

3 **Jerome Kern** *and Jean Harlow pose on the set of* <u>Reckless,</u> *1935. The original photo caption notes that Harlow gave Kern's song ''a game try.''*

2

3

31

composer may be consulted during the early phases of production as the director develops his ideas for the movie's style, but composition does not begin until the film has been edited. The composer views the film with the producer and/or director and together they determine the general direction of the score. The composer will also work closely with the music editor, who often provides creative input and supplies the composer with precise timings of the scenes to be scored.

Composing for film is very precise – the length of musical passages is sometimes measured to the fraction of a second – and the composer has the film and viewing equipment with him throughout the process. The composer may also serve as the conductor on a specially equipped soundstage where the film is screened as the musicians play.

1 **Nacio Herb Brown** (seated) and **Arthur Freed** (later a producer) with Bessie Love and Anita Page, at work on The Broadway Melody, 1929.

Concessions Worker

Concession counters came into being when the Depression began to shrink box-office proceeds and concessions workers have kept the popcorn flowing ever since. Because concessions revenues can mean the difference between profit and loss for theater owners, concessions workers are expected to keep a strict accounting of cash intake and in some instances even tally the number of beverage and popcorn cups with the day's sales.

1

The conductor of a film score is often the composer as well, and his job is to lead the musicians to a suitable interpretation of the score. His work begins late in post-production, and he must thoroughly familiarize himself with the score and the editor's approved work print, confer with the director for his thoughts on the score's proper interpretation, and conduct rehearsals.

The music is recorded on a sound stage with screening facilities; there the conductor views the film as the music is recorded to match the action on the screen. When the recorded score is mixed with the other sound tracks, film and sound are ready for final processing in the laboratory. (AFM)

2 **Dimitri Tiomkin** *conducting his score for* Duel in the Sun, *1946. Up on the screen are Jennifer Jones and Gregory Peck.*

3 **Fred Waring** *leading his orchestra at the CBS television studio, 1952.*

2

3

Construction Coordinator

The construction coordinator is responsible to the production designer or art director for building all the sets, flats, backdrops, and sometimes furniture required by a film or television production. Keeping one eye always on the budget, the coordinator plans a construction schedule and selects an appropriate crew. The construction department works from the drawings and plans submitted by the art director, and construction begins well before the start of shooting.

On a film production, construction often continues into shooting, since sets needed later in the schedule are built while others are in use. Construction coordinators building for television often work for a network or independent scenic shop and generally work on small-scale projects with tight budgets and schedules. Construction coordinators must have considerable experience in carpentry as well as acute managerial skills. (IATSE, NABET)

1 **Harold Fenton** *with the construction schedules for* Gone with the Wind, Intermezzo, *and* Rebecca, 1939.

1

Construction Foreman

On productions involving a great deal of construction, especially when work is under way in more than one place at the same time, construction foremen are assigned to supervise work at each site. For example, if construction occurs simultaneously on a location and in the studio, the construction crew's day-to-day activities are supervised by a foreman, allowing the construction coordinator to assume a more administrative role. The foreman probably receives direction from the production designer and art director but is, of course, responsible to the construction coordinator. (IATSE, NABET)

The costume designer outfits principal actors, bit players, stunt performers, and sometimes extras, with any combination of original designs, purchased clothing, or rented items from costume companies. The costume designer begins work well in advance of shooting, meeting with the director or producer to determine such things as the period, nuances of social class, and the relationship between costume and character. The process usually involves research, especially on a period film, but even a contemporary production may require an investigation of the dressing habits of a particular place or group of people. The costume designer also consults with the production designer and the director of photography, since set design and the production's photographic style often influence the choice of pattern, texture, and color. On taped television productions the costume designer may cultivate a relationship with the video operator, whose manipulation of the video image also has an effect on the appearance of clothes.

The designer prepares a breakdown of all

2

2 **Irene** makes last minute adjustments to Irene Dunne's veil before the wedding scene in The White Cliffs of Dover, 1944.

3 **Walter Plunkett** sketching Scarlette O'Hara's '' curtain dress'' for Gone with the Wind, 1939.

3

costumes and accessories from the script, then makes preliminary color sketches and fabric selections for the director or producer's approval. Once approved, the designer arranges for original costumes to be made (usually by a costume maker at a costume shop though he or she may hire costume makers and set up shop) and purchases or rents others.

Costume accessories – jewelry, shoes, gloves, handbags – are also within the designer's domain, although some incidental items, such as watches and umbrellas, belong to the property master and require coordination between the two departments.

Film and television productions present many special problems that the designer must take into account: the sound a fabric makes as the actor moves about; the need for several identical costumes in the likely event that a stunt requires more than one take; and the aging or soiling of costumes with paints, stains, dyes, or even files and rasps.

The designer is usually present on the set throughout shooting to supervise all costume work and to handle any emergencies or last minute changes. He or she also works closely with the makeup artist and hairstylist to ensure coordination between all aspects of a performer's appearance. (IATSE, NABET, USA)

2

1 **Adrian** reviewing designs at MGM, 1930.
2 **Edith Head** shows her designs to Danny Kaye and Mai Zetterling, stars of Knock On Wood, 1954. Writer Norman Panama looks on.
3 **Helen Rose** makes a final check of her design for Lena Horne in Till the Clouds Roll By, 1946.
4 **Travis Banton** (seated second from left) and **Edith Head** (seated at right) watch Mae West show off one of their designs for Klondike Annie, 1936.

1

3

Assistant Costume Designer

In a complex film or television production the costume designer has an assistant to help with any responsibilities that are not handled by the wardrobe department. These include preliminary costume sketches, color and fabric selections, shopping and rental, and any special preparations required, such as aging or staining the costumes. (IATSE, NABET, USA)

4

Costume Maker

Costume makers build original designs and special costumes from the designer's sketches, custom-fitting them to the actors during a series of fittings. Working from the designer's color sketches and the actors' measurements, the costume maker creates the patterns, orders all necessary materials, and may also conduct research into the period or setting of the film.

Besides constructing original designs, costume makers often duplicate bought or rented costumes if there are scenes in which clothes will be soiled or ruined. On most productions, the costume maker supervises a staff of skilled seamsters or seamstresses. Costume making may continue into the shooting phase, but the costume maker is rarely on the set; the completed costumes become the charge of the wardrobe supervisor during shooting. (IATSE, USA)

1 *It took more than twenty dressmakers to build the hoop skirts for* Gone with the Wind, *1939. This one was bound for Vivien Leigh.*

2 **Arthur Levy** *and* **Helen Wood** *with dress forms of the stars at 20th Century Fox, 1938.*

1

2

Crane Operator/Grip

Crane operators are responsible to the key grip for running the camera cranes — huge mobile machines with hydraulic arms that move both vertically and horizontally. As many as four grips may be needed to drive a crane and adjust its arm for each shot. Operators practice crane movements during rehearsals and work closely with the camera and electrical crews.

The cranes are massive and potentially dangerous, and they suspend the camera operator and the director of photography dozens of feet in the air. As a result, the crane crew must be skilled mechanics as well as operators, maintaining their equipment at the highest standards of safety. (IATSE, NABET)

3 *Duel in the Sun,* 1946.
4 *Xanadu,* 1980.

4

3

DGA Trainee

A two-year on-the-job training program designed to teach the responsibilities of the assistant director is offered by the Directors Guild of America. Candidates for the program must apply to the Guild and are evaluated through a highly competitive selection process involving a written examination, interactive group evaluations, and interviews. The Guild assigns trainees to particular productions. On the set they work under the second assistant director's supervision. Upon successful completion of the program, trainees gain full union membership as second assistant directors.

. . . .

Dialogue Coach

Productions require a dialogue coach when an actor or actors must speak with a foreign accent or in a regional dialect. To prepare an actor for such a role, a dialogue coach may use tape recordings, plan research trips, and provide private lessons as well as coaching on the set.

Dialogue directors were relatively common in Hollywood after the switch from silent films to talkies. Many silent film directors had no experience coaching actors with their lines and stage directors, often from Broadway, were recruited to fill the void. As the sound film established itself, the need for the two directors gradually disappeared.

Of all the people who work behind the camera, only the director has managed to capture the public's imagination as much as the actor. Stereotypes abound — the bellowing tyrant, the "actor's director," the auteur, the hack — and indeed there are as many different styles of directing as there are different kinds of scripts. A director adept at drama may be hopelessly lost with an action-

1 **Sidney Lumet**, _Stage Struck_, 1958.
2 **Cecil B. DeMille** with Charlton Heston, _The Ten Commandments_, 1956.
3 **George Cukor** with Katharine Hepburn, _The Philadelphia Story_, 1940.

1

2

3

packed adventure, and vice versa. Regardless of the genre, the director is the creative center of a production: He shapes and guides the creative contributions of dozens of artists and technicians, from writer to re-recording mixer, and bears artistic responsibility for the result.

Ideally, the producer hires the director at or near the project's inception. The director first attends to the script, either writing or rewriting it himself, or working with one or more writers until he is satisfied with the result. Though the producer hires the majority of crew members, the director usually chooses key creative personnel, such as the director of photography and production designer, and is very involved in casting. The director who

1 **Dorothy Arzner** *with Helen Percey, head of Paramount's Research Department,* Merrily We Go to Hell, *1932.*

2 **Mervyn LeRoy**, Waterloo Bridge, *1940.*

3 **D.W. Griffith** *and Cameraman Billy Bitzer, 1914.*

4 **Frank Capra** *with Claudette Colbert and Clark Gable,* It Happened One Night, *1934.*

5 **Elia Kazan** *with Karl Malden and Vivien Leigh,* A Streetcar Named Desire, *1951.*

6 **Susan Seidelman**, Smithereens, *1982.*

7 **King Vidor** *(second from left) with Producer Irving Thalberg and Renée Adorée,* The Big Parade, *1925.*

8 **John Ford** *with Tim Holt,* Stagecoach, *1939.*

1

3

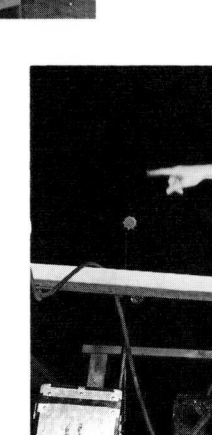

2

4

5

6

7

8

takes an active role in all pre-production processes, from costume design to location scouting to special effects, wields great creative influence from the start. However, this is also the time when he will have to make the most compromises, adapting his vision to the limitations of time, budget, and physical possibility.

On the set the director's control is more or less complete. During shooting, he works most closely with the actors and the director of photography. He coaches the actors, rehearsing them if he chooses, and determines lighting and camera placement and movement with the director of photography. In multiple-camera video productions, the director makes the camera decisions himself and relies on the camera operators to execute them.

Though the Hollywood director of the thirties and forties often saw no more of the film when shooting ended, today's director is usually involved with every area of post-production: composing, sound editing, music editing, re-recording, and especially editing.

It should be noted as well that the decentralization of film production after the disruption of the studio system in the fifties has given rise to a greater number of filmmakers who assume multiple roles, such as director-producer, writer-director, writer-producer, and even writer-director-producer. In these instances, the individual may approach a point of almost total artistic control, but the collaborative nature of film and television assures that creative contributions are never limited to any one person. **(DGA)**

1 **Peter Bogdanovich**, _Nickelodeon_, 1976.

2 **Billy Wilder**, _Irma la Douce_, 1963.

3 **Georg Stanford Brown** (center) on the set of the television movie _"Grambling's White Tiger,"_ 1981.

2

1

3

44

The director of photography is a skilled cameraman. The director relies upon his extensive knowledge of lighting, optical principles, and camera equipment to put the story on film with the desired effects of mood, color, composition, and movement.

The choice of a director of photography is one of the first decisions made by the director and producer, and here the producer often defers to the director's preference. The director of photography takes an active part in preliminary production conferences, offering suggestions about the budget and the choice of film stock, conferring with the production designer about the plans for sets and locations, informing the production manager of the camera and lighting equipment required, and assembling the camera crew. He maintains these working relationships throughout production, but his closest collaboration is with the director, who is the final authority on the film's visual interpretation.

When shooting begins, the director of photography's authority on the set is second only to the director's. Together they determine

4 **James Wong Howe** checks out a camera angle on the set of The Hard Way, 1942.

5 **Carlo DiPalma** (second from right) and Director Woody Allen line up a shot for Radio Days, 1987.

5

4

45

where to put the camera, how it will move, and what lighting is appropriate for each scene. It is a constant barrage of choices to make, each choice creating and defining the next.

The director of photography gives the gaffer his instructions for lighting and directs the camera crew to the proper setup. Camera rehearsals follow, conducted either with the actors or with stand-ins, allowing the director of photography to check the gaffer's work and lead the camera crew through a dry run of all camera angles, movements, and focus adjustments. The sound crew often participates as well, for coordination between sound and camera equipment is essential.

The director of photography carries a small viewfinder fitted with a variety of lenses to evaluate preliminary and potential setups, and he always looks through the camera beforehand to check the composition. He may also supervise an array of specialized camera equipment, such as crane-mounted and remote-control cameras. Although special-effects teams often do their own photography, he must oversee it to ensure visual consistency. The director of photography views the rushes with the other key members of the production team, providing a chance to evaluate techniques and make suggestions to

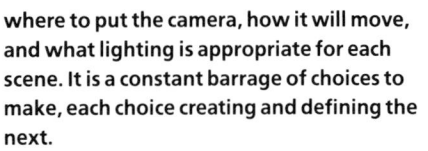

1

2

the director. He also instructs the film laboratory on which of the wide range of printing
techniques and processes should be
employed to achieve the desired appearance
of the film.

Becoming a director of photography usually involves years of experience on the camera crew, from clapper/loader to focus puller
to operator, and, finally, the top camera position. Directors of photography maintain a
strong professional association – the American Society of Cinematographers – which
publishes its own magazine and disseminates
information about new techniques and tech-
nologies. (IATSE, NABET)

1 **Giuseppe Rotunno** and Director Bob Fosse confer
on the set of All That Jazz, 1979.
2 **Charles Rosher** takes a light reading on the set of
Kismet, 1944, as Marlene Dietrich talks with her
horse, Reno.
3 **Ray Rennahan** measures the light intensity over
Ingrid Bergman and Gary Cooper on the set of For
Whom the Bells Tolls, 1943.
4 **Michael Chapman** (at camera) checks a shot of
Robert DeNiro for Raging Bull, 1981.

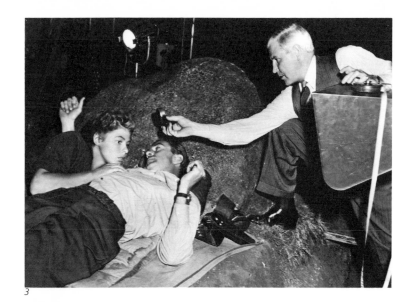

3

4

Dolly Grip

By mounting the camera on a dolly that runs along tracks, many otherwise difficult moving camera shots can be achieved. The dolly grip is the person responsible for laying the tracks and pushing or driving the dolly to achieve the desired effect.

Dolly shots are often very complex, requiring the dolly grip to work closely with the camera crew to perfect the movements during rehearsals. During moving shots, dolly grips may also push the wheeled platform holding the microphone and boom operator. **(IATSE, NABET)**

1 Pushing camera and microphone dollies around a distraught Vivien Leigh, Gone with the Wind, 1939.

1

Draftsperson

During pre-production and often into shooting, draftspeople are engaged by the production designer and art director to execute the working drawings, renderings, plans, and elevations from which sets are constructed. A draftsperson may be called upon to draw anything from an entire set to small portions of decorative detail. **(IATSE, NABET, USA)**

2 Working on sets for The Broadway Melody, 1940. Production Manager Art Smith (at left) looks things over.

2

The driver receives his orders from the transportation captain, who oversees the transporting of film and television equipment, scenery, and personnel during shooting. He may drive a large rig cross-country, a motor-home dressing room to location, or chauffeur directors and actors to and from the studio or location. He may also drive cars rigged up to hold the camera for moving shots. **(IBT, NABET)**

3 Driving Director Cecil B. DeMille (white shirt in passenger seat) and the camera crew for a scene in The Ten Commandments, 1923.

3

The electrician works under the supervision of both the gaffer and best boy during shooting. There are usually several electricians on the set or location and they do whatever is necessary to achieve the lighting plan devised by the director of photography, including position cables, set up and adjust lights, and install wiring. (IATSE, NABET)

1 Focusing a light on Marion Nixon for Rebecca of Sunnybrook Farm, 1938.
2 Positioning lights over Whitney Bourne, Claude Rains, and Director Charles MacArthur on the set of Crime Without Passion, 1934. Director of Photography Lee Garmes sits at left.
3 Four of the huge arc lights used for The Crusades, 1935. The original photo caption states that the film used enough light ''to illuminate a town of 25,000.''

3

1

2

Executive Producer

The ideal executive producer is a businessman with a good instinct for commercially viable material, and a knack for assembling a creative team that will deliver the best possible result.

However, the executive producer is usually most involved with only the initial stages of a production – creating or finding a suitable concept, story idea, or property and selling it to an organization, for example, a studio or network, or individuals who will finance its production. To make the project more attractive to potential backers, he may put together a "package" of a property, director, and stars.

The executive producer often retains a "piece" of the production, meaning a percentage of any profit it makes. Though executive producers, especially in television, may assume a strong creative and administrative position throughout a production, many never step on the set.

Extra

All those people in the background – milling about in crowds, having lunch in restaurants, dancing at fancy dress balls, dashing around busy offices – don't just suddenly materialize when the director calls for action. Extras, sometimes by the hundreds, fill in the background and provide the human element supporting the illusion that the lead actors are real people functioning in a real environment.

Extras may do very little or they may be sought out for their special abilities, such as horseback riding or ballroom dancing. The responsibility for finding and hiring extras usually falls to the producers or assistant director, who directs all background action. (AFTRA, SAG, SEG)

4 Storming the casting office door at the Astoria Studio, hopefuls try to get on the set of Fine Manners, 1926.

4

A casting company, usually not the same company that cast the principal actors, is hired to locate and cast extras. The extra casting director usually meets with the director at the beginning of production to discuss general guidelines, but works most closely with the first or second assistant director, who provides the extra breakdown list with specific instructions regarding requirements for age, gender, special abilities, etc.

The extra casting director finds the extras through his own talent files or advertises an open call in the trade papers. When searching for extras on a remote location, he may advertise in local papers or on the radio.

The extra casting director works through all or part of shooting, usually on very short notice, getting breakdowns just a day before the extras are needed.

The planning and shooting phases of a film production represent a massive investment in time, money, and the creative efforts of sometimes hundreds of people. However, the result of this collective effort is essentially formless: hundreds or thousands of shots of varying lengths, taken from a variety of angles and usually out of narrative sequence. The film editor shapes this celluloid tangle into an effective dramatic form, bringing the film to life as it is envisioned by the writer, director, producer, and the editor himself, whose fresh perspective and creative sensibility play a significant part in what will eventually appear on screen. The director/editor working relationship is often one of the closest in film production, and many directors have moved into the director's chair from the editing room.

The film editor begins his work as soon as the lab processes the footage from the first day's shooting. He and his staff use a synchronizer to match the images with the sound, cut full reels into individual takes, keep extensive written records, and carefully

1 **Danny Gray** and **Blanche Sewell** *going over footage from* Queen Christina, *1933.*

2 **Anita Posner**, *1975.*

3 **Jordan Leondopoulous** *editing* Wolfen, *1981.*

monitor both picture and sound for consistency and quality.

Each day, the editor, director, and producer view the previous day's footage and together they choose the takes that will be used. At the conclusion of shooting the editor assembles these takes into the work print, a rough draft of the film. Using an editing machine that allows him to view segments of the film repeatedly and at a number of speeds, he cuts and splices, experiments with different sequences of shots, and trims and molds the film until he and the director are satisfied. He marks the film with a grease pencil to indicate optical effects, and prepares detailed instructions for the lab and the negative cutter. He also works closely with the sound editor, often acting as liaison between sound editor and director, and contributes to the planning of the sound track. (IATSE, NABET)

1 **Jane Loring** in an editing room at Paramount.

2 **Viola Lawrence** editing First Comes Courage, 1943.

1

2

Apprentice Film Editor

Responsible to either the assistant film editor or the film editor, the apprentice film editor is learning the ropes and paying his dues. As part of his apprenticeship, he may perform a variety of tasks, from filing to keeping the editing room in order to running errands. Since his primary objective is to learn and observe, he is at the disposal of the editing team and may do some work with them on a limited and closely supervised basis. (IATSE, NABET)

Assistant Film Editor

The important and creative business of editing also involves many time-consuming and laborious tasks that fall to the assistant editor. These include synchronizing picture and sound tracks for the viewing of dailies, breaking down the dailies into individual takes, filing and cataloguing massive amounts of footage, communicating with the laboratory, ordering supplies, and cleaning and maintaining all film, sound tracks, and editing equipment. By assuming the responsibility for many of these duties, the assistant editor frees the editor to concentrate on the actual creative editing.

The assistant also accompanies and assists the editor at meetings, screenings, and other sessions, bringing all necessary materials and taking notes as required. Under the editor's supervision, the assistant will usually have the chance to do some editing as well, since the position is considered preparatory to becoming a full-fledged editor. (IATSE, NABET)

First Assistant Camera Operator (Focus Puller)

The first assistant camera operator is called the "focus puller" because he moves the focus ring on the camera lens as the distance between the camera and the object that must remain in focus change. Following the instructions of the director of photography, he fits the camera with the correct lenses, sets the aperture, attaches filters if needed, and must keep the camera loaded and clean. He uses a tape measure to determine the distance from the lens to the point or points of focus; then records the distances in a log that he attaches to the camera for reference.

The assistant operator cannot see the results of the focus adjustments — he is never looking through the camera — and instead relies on the measurements he takes and figures engraved on the lens to achieve the proper focus at each distance. Before shooting begins these camera adjustments are thoroughly practiced. The focus puller may also be responsible for loading the film into the camera, threading it, and for checking that the interior is clean and free of dust.

The first assistant has usually had experience as a second assistant, and must be knowledgeable enough to operate the camera effectively in an emergency. The first assistant's logical path of advancement is to camera operator and, with talent and a little luck, to director of photography. (IATSE, NABET)

1 Measuring the distance from lens to Leslie Howard for Intermezzo, 1939.
2 Filming costume tests for 1776, 1972.
3 Adjusting focus during a scene with Peter Lorre and Joan Lorring, Three Strangers, 1946.
4 An unusual dolly shot for New York, New York, 1977.

2

1

3

4

The sound track recorded during filming rarely contains all the sound effects needed for the finished production. The sound editor turns to the Foley artist to create sound effects tracks to supplement the production and library sound tracks. In a specially equipped studio, portions of the film are screened along with their existing sound accompaniment, and the Foley artist creates new sounds in the studio to match their source on film.

Much of the the Foley artist's work is the creation of small effects such as footsteps and body rustles. However, by layering as many as twenty individual tracks, he can create large-scale sound effects such as car crashes or the demolition of a building. He has a prop room full of materials and implements to make the sounds, and supervises a recording engineer who monitors and records them.

Gaffer

The gaffer is the film crew's chief electrician, and it is his responsibility to coordinate and supervise the lighting on a set or location. He works for the director of photography, who determines the lighting plan and informs the gaffer of the requirements for each shooting day in advance.

Lighting plans may be very complex, with lights hung from above, mounted outside windows, and raised on scaffolding, and the gaffer must ensure that none of the equipment falls within camera range. The gaffer supervises a staff of electricians who set up and operate the equipment, and they participate in all camera rehearsals to test the effectiveness of the lighting plan. (IATSE, NABET)

1 *Positioning lights above the set of* Sally, Irene and Mary, *1925.*

1

. . . .

Generator Operator

When portable generating equipment is required to run cameras and lights on location, a generator operator is engaged. Directly responsible to the gaffer, he remains at the generator at all times to ensure its correct output. (IATSE, NABET)

Grips provide the labor intensive skills required by a business that may use ten tons of equipment for every scene put on film. The grip is responsible for loading and unloading equipment, moving and positioning sets, scenery, and backdrops, erecting scaffolding or platforms for camera and lighting equipment, keeping cables free of interference during moving shots, repairing minor damage to sets, and even devising ways to attach cameras to moving objects such as cars or helicopters. Responsible to the key grip, he may also assist the dolly and crane grips. (IATSE, NABET)

2 *Wheeling in sound equipment in preparation for shooting.*

2

59

Hairstylist

The hairstylist attends to all the hairdressing needs – cutting, styling, dyeing, wigs, etc. – of the performers on film and television productions. Since hairstyles help establish the character and mood of the performers, the hairstylist works closely with the director, costume designer, makeup artist, and possibly the production designer or producer.

On film productions, the hairstylist usually starts work just before shooting begins, however, a period production may require lead time to conduct research on hairstyles of the particular era. In television, the hairstylist may be on the staff of the company or network producing the show, in which case he or she will start a new production weekly or daily. During filming or taping, the hairstylist works most closely with the costume designer and makeup artist; each contributes to the appearance of the performer, who must emerge as a believable screen character. Hairstylists must be present at all times for touch-ups or repairs between takes. On very large productions, he or she will have assistants who are skilled hairstylists in their own right, but whose work is coordinated and overseen by the head stylist. (IATSE, NABET)

1 Jack Red Bear gets a touch-up on location for _They Died With their Boots On_, 1942.

2 **Sidney Guilaroff** fixes up an improbable hairdo for Marlene Dietrich, _Kismet_, 1944.

1

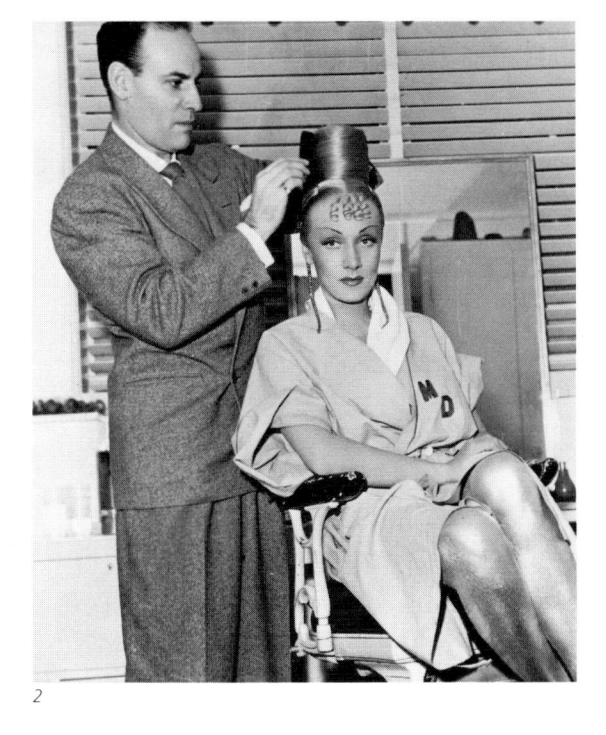

2

60

Well before shooting begins, the key grip meets with the director of photography, to whom he is responsible, to determine the production's needs in grip equipment and personnel. During rehearsals and shooting he coordinates and supervises the work of all grip personnel, who perform a range of labor intensive tasks, and often works together with them. Though the key grip works most closely with the camera and lighting crews, it is likely that he has at some point worked with and assisted virtually everyone on the set. (IATSE, NABET)

Many lab technicians work on a film project from the day shooting begins to the final preparation of prints for release. Their work entails a variety of technical skills that ultimately affects the quality the film seen in theaters.

When negative comes into the lab, it is scanned for damage, developed, then closely inspected by technicians for any irregularities or damage. The color timer examines the film's tones and values and prepares instructions on how the film should be printed. The resulting prints are inspected and cleaned again before being developed and the film is then screened to determine if its quality is acceptable. It is then either sent back for correction or placed on reels and shipped to the customer.

At each stage along the way, lab technicians work with film that is extremely valuable and delicate — a roll of negative represents an enormous expenditure of time and money — and they must handle it with the greatest care. Lab technicians usually begin as all-round assistants, training almost exclusively on the job. With talent and experience the technician may advance to the top technician's position of color timer. (IATSE, NABET)

1 **Adele Simonsen** *in the printing department at Movielab, 1975.*

2 *Printing at the Warner Bros. laboratory, c. 1952.*

1

2

Lighting Director

For a television production, the lighting director plans and implements all lighting arrangements on the set. He studies the script to determine the requirements of the show — for example, the action, mood, and time of day — and, as the show develops, plans the lighting with careful attention to camera placement and movement. The lighting director makes a major contribution to the quality of the image and therefore he must work closely with the camera operators and the video operator to achieve a cohesive and attractive picture.

Lighting for television has fundamentally different requirements than lighting for film. Film lighting is most often designed around the position and angle of one camera, whereas in television a lighting plan frequently must be equally effective for each of four or more cameras in a wide range of positions on the set. (IATSE, IBEW, NABET)

3 The lighting switchboard at NBC's Color City studio in Burbank, California, 1957.

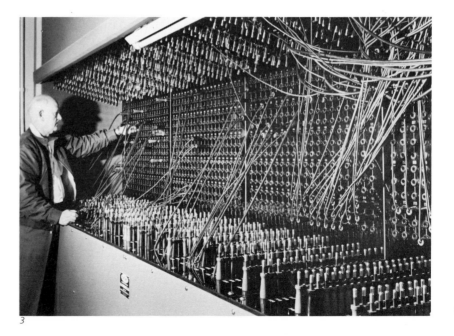

3

The producer or production manager hires a location manager to coordinate the planning and administrative details of shooting on location. Ultimately responsible to the producer and director, the location manager works most closely with the production manager. After reading the script, the location manager usually meets with the director, producer, production manager, production designer, and other department heads to discuss location requirements and possible sites. If the script calls for a scene in Grand Central Station, location scouting is rather uncomplicated, but most of the locales described in scripts are not specific, and the location manager develops three to five possibilities for each location and hires scouts to assist in documenting them.

Once selected, the location manager must make arrangements with the property owners for use of the site. This may involve negotiating fees, approaching city governments and other local authorities for permissions, and filing permits. He or she must also make sure that property owners are aware and approve of the activities of the production company. For example, the art department may plan to significantly alter the appearance of buildings or interiors and, understandably, property owners may consent to the changes only on the condition that the sites will be restored to their original appearance.

The location manager also arranges for all support services needed during location shooting – catering, changing rooms, tents, makeup trailers, portable bathrooms, etc. He or she may have a sizable staff, often composed of location scouts who remain on the production as necessary to serve as assistants. (DGA)

1 **Lou Strohm** *maps out the transportation route for location shooting of* The Bad Man, *1941.*

1

Location Scout

If assistance is needed in finding appropriate locations, the location manager hires one or more location scouts. The director, production designer, and location manager develop lists of possible locations; the scout is then sent out to assemble extensive photographic documentation of the sites under consideration and to gather any other relevant information. The director or producer makes final location choices based on the information provided by the scout.

· ·

Louma Crane Operator

The Louma crane, similar in appearance to a microphone boom, is a modular, robotic crane system that allows lightweight film or video cameras to be operated from a "ground station" equipped with a video monitor displaying the image seen by the camera. By freeing the camera from the operator, the Louma crane allows for previously impossible camera positions and movements. The Louma crane can be mounted on virtually anything — camera cars, elevated platforms, balconies, even rafts.

When a Louma crane is needed, the operator meets with the director and director of photography to plan its use. The operator designs a system to achieve the desired shots and, on the set or location, sets up the equipment, oversees its use, and sometimes operates the camera controls during the shot. In general, however, camera movement and focus adjustment are handled by the production's camera operator and focus puller. (IATSE)

2 Using the Louma crane on location for Ragtime, 1981.

2

Makeup Artist

The makeup artist is responsible for the actors' cosmetic requirements on film and television productions. His duties range from straightforward, contemporary looks to making an eighty-year-old man out of a thirty-year-old actor, or a Mr. Hyde out of a Dr. Jekyll.

On film productions, the makeup artist often receives a script well in advance of shooting, then consults with the director, costume designer, director of photography, and production designer to plan the makeup and ensure its integration with the visual content of the film as a whole. In television shows, such as sit-coms, talk shows, and news broadcasts, the makeup artist is often on the staff of the network or production company, and his primary task is to create natural, contemporary looks for performers and guests to compensate for the negative effects of harsh lighting and the camera's ability to magnify flaws and blemishes.

The makeup artist works most closely with the costume designer and hairstylist, who are also contributing to the appearance of individual performers, and their work must be integrated. Besides natural makeup styles, however, the makeup artist has a bag of tricks with which he can create many spectacular effects. Using materials such as latex, rubber,

1

cloth, baldcaps, prosthetics, and "blood," he can create wizened historical characters, monsters, Martians, or simulate sores, bruises, and any number of wounds and injuries. When productions require extensive special effects or elaborate character transformations, such as fake heads and limbs, a prosthetic makeup specialist is called in.

Like the costume designer and hairstylist, the makeup artist may conduct research to create a period style. He is present on the set or location at all times, where in addition to applying makeup he must be available for touch-ups between takes and makeup removal. (IATSE, NABET)

1 **George Fiala** *makes up Richard Waring as Aaron Burr for an episode of CBS's "You Are There," 1954.*

2 *Paul Robeson getting a touch-up on the set of* The Emperor Jones, *1933.*

3 **Perc Westmore** *turns Paul Muni into Emile Zola for* The Life of Emile Zola, *1937.*

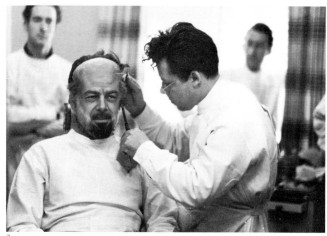

3

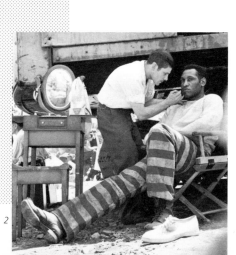

2

67

The exterior of Charles Foster Kane's enormous mansion Xanadu in <u>Citizen Kane</u>, 1941, was not an actual set or house but a matte painting. Matte painting is a time-honored film technique that involves combining painted backgrounds with live action, and it is still very much in use today. The technique is often used to add scope to the appearance of a building too expensive to build: The actors can work in front of a set representing the building's first story; the upper stories, towers, and surrounding atmosphere can be painted in. It can also be used to eliminate expensive location trips – a painting of the Alps can be combined with footage shot on a California backlot – or to take the production to mythical locations, such as Mars or an underwater city.

The matte painter works with the director and production designer to develop a concept and to determine the painting's relation to the live action. The live action portion of the scene is always shot first, and the matte painter is usually on the set when it is shot to help him visualize the completed scene. He then takes the footage to his studio, where he projects a frame of the live action to determine the area to be painted. The painted and live action portions of the scene are photographed and printed together, and fit like two pieces of a jigsaw puzzle. The matte painter must be able to paint everything from architecture to moonscapes and at the same time be sufficiently familiar with camera and lighting techniques to achieve a perfect blend of illusion and reality. (IATSE)

1 **Albert Whitlock** with a matte painting he created for the mini series "Masada," 1981.

1

Mechanical Effects Designer

Mechanical effects is a broad term embracing a wide range of special effects. The mechanical effects designer may create convincing animals, monsters, or human limbs, rig up cars to create sticky situations for stunt performers, or use wires to make people or objects appear to fly.

During pre-production, he will usually meet with the director and producer to discuss the script, plan the effects, and determine a budget. The designer may have considerable creative input at this stage because the director may know what he wants but hasn't the foggiest how to achieve it. Instead the director relies on the designer's knowledge and ability to combine sophisticated artistic and technical skills. For example, mechanical animals are usually composed of a realistic outer shell concealing a maze-like inner framework containing the cable and hydraulic components that will bring the animal to "life" when manipulated by the designer and his crew.

Flying people and objects present a different set of problems: The designer creates a system by which an actor or object is suspended by wires from an overhead apparatus that utilizes tracks, pulleys, or hydraulic equipment to simulate flight. The mechanical effects expert is called upon to adapt cars for effects or stunts, such as selectively sabotaging a vehicle for a planned crash. In such cases, the designer works closely with the stunt coordinator and may outfit the car with safety cages or other special equipment to ensure the stunt performer's safety.

The designer often has a large staff to aid in the design and construction of equipment and its operation. (IATSE, NABET)

2 **Eoin Sprott** with one of the mechanical wolves he created for Wolfen, 1981.

2

Miniatures, which fall into the realm of special effects, have been a valuable tool for filmmakers for decades. Film scripts often call for the depiction of events that would either be impractical or impossible to create at full scale, such as the explosion of a gigantic spaceship or the destruction of downtown Los Angeles by an earthquake. Both types of effects, however, can be achieved very realistically through the use of miniatures.

Miniatures can be used for sea and air battles, shipwrecks, floods, and other natural and man-made disasters. The producer calls in a miniature specialist when any such work is required.

Sometimes the miniature specialist designs and builds the models; often he builds them from the production designer's specifications. Whatever the arrangement, extensive miniature work involves extensive communication with both the director and the production designer, who often builds full-scale sets for the actors to work in meant to represent a portion of the miniature. When full-scale and miniature footage are combined, the two must flow together seamlessly.

The miniature designer usually has a staff of model makers, who construct the models

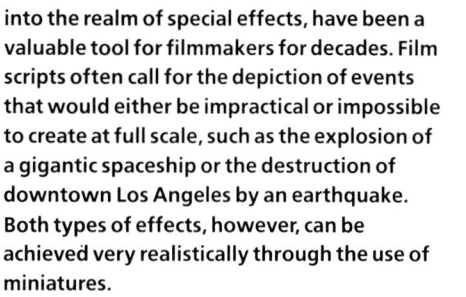

1

out of any number of materials – spare parts culled from model airplane kits, molded plastic, latex, and more conventional materials. Miniatures are usually built as large as possible since this allows for the necessary level of detail and because the camera is unable to convince us of the reality of miniatures smaller than one-sixteenth scale.

Indeed, the camera plays a very large role in the success of these illusions, and the miniature designer must therefore coordinate his work with the director of photography. Miniatures are filmed at faster than normal camera speeds in direct relation to the scale of the model. For example, the explosion of a quarter scale miniature boat filmed at twice normal camera speed appears twice its actual size. (IATSE, NABET)

1 Working on a scale model of Manhattan for Invasion, U.S.A., 1952.

2 Positioning a rather large miniature for Raise the Titanic!, 1980.

2

Model Maker

Model makers, possibly very many of them, work for the miniature supervisor during pre-production and shooting. The model maker is an expert craftsman whose job has nothing if not variety: He and his colleagues may create miniature boats, planes, cars, spaceships, buildings, or whole landscapes, complete with trees and mountains, in which the miniatures can be manipulated and photographed for realistic effect.

The model maker builds models from plans created by the miniature supervisor or production designer, and the job demands skills ranging from metal-working to sculpture to painting. He is often present on the set to operate the models – planes or spaceships flown on wires, baking-soda snow falling on a miniature forest – and to perform touch-ups and repairs between takes. (IATSE, NABET)

Music Arranger

Film and television productions are as likely to use existing music as original music, and everything from Beethoven to Cole Porter can find its way into a sound track. During post-production planning of the musical score, the music arranger, who in many cases is also the composer, conductor, or both, adapts existing music to suit the production, re-scoring it as needed.

The arranger may work with the director, producer, film editor, or music editor to determine the precise passages with which he is to work. Since he must refer to specific scenes, he views a print of the film while working on the arrangements. (AFM)

Music Editor

When the film editor completes the final cut of the film, the music editor begins his work by meeting with the director, composer, and film editor to "spot" the film – to determine where music will be, which moments it will highlight, and the mood or atmosphere it should create.

The music editor acts as a liaison with the composer and the director, producer, and film editor; supervises recording sessions; and edits the music tracks to match the picture. He assists the composer by preparing precise timings of the scenes that need to be scored and, since he usually has an extensive musical background, may have considerable input into the composition. He works with the recording engineers in the booth during recording, and then takes the tracks to the cutting room to build the final music track. The music editor also prepares cue sheets for the re-recording mixers and attends mixing sessions. (IATSE, NABET)

Assistant Music Editor

The assistant music editor works for the music editor late in post-production. He is responsible for maintaining the editing room, synchronizing music and picture tracks, filing and cataloguing the tracks, and may also assist in the preparation of timings for the composer and cue sheets for the re-recording mixer. (IATSE, NABET)

Music Recording Supervisor

The primary responsibility of the music recording supervisor is to see that the music is properly recorded and to act as a coordinator for the composer/conductor and the director, producer, or film editor. He may also hire musicians and schedule recording sessions. Not every production requires a music recording supervisor and some of the job's responsibilities may be assumed by the film or music editor. (AFM)

Musician

Although musicians rarely get screen credit, the quality of musical scores owes more than a little to the professional musicians who work in film and television. Hired by a contractor or the music-recording supervisor, musicians play for the conductor during all rehearsals and recording sessions. While most musicians are unknown to the public, a few familiar musicians play on screen for television shows with their own resident bands. (AFM)

The Philharmonia Orchestra of London records the score of Hamlet, *1948, under the gaze of Laurence Olivier.*

Musician Contractor

A musician contractor is hired to assemble the required group of musicians and to act as the orchestra manager for the recording of the score during post-production. Responsible to the music editor or the composer/conductor, the contractor is present at all recording sessions and is often an instrumentalist as well. The contractor may also be called upon to assemble a group of sideline musicians to work during shooting. (AFM)

Negative Cutter

The negative cutter is responsible for one of the last crucial steps in the production process. When film and sound editing are completed, the film editor gives the final, approved work print – the spliced, trimmed, and edited dailies put together as a complete film as it will be seen by the audience – to the negative cutter, who cuts the original negative to match it frame for frame. This is an extremely delicate process and the fragile and valuable negative must be handled with great care and precision. Using a splicer, the negative cutter refers to the edge numbers printed on the sides of both positive and negative film and follows detailed instructions from the editor. The cut negative is then sent to the lab to be printed in quantity and combined with the completed sound track for release. The negative cutter is supervised by the film editor, and usually works in the cutting room rather than the lab. (IATSE, NABET)

1 **Lee Fontana** *working at the Astoria Studio, c.1950.*

1

Studios and networks often maintain a trained medical staff to handle any minor injuries sustained by cast and crew members. When potentially dangerous stunts or explosive effects are performed, the production company must arrange for an ambulance and paramedics to be on hand. (IATSE, NABET)

2 **McWilliams** *and nurse attend to a minor injury at the Fox Hills Studio.*

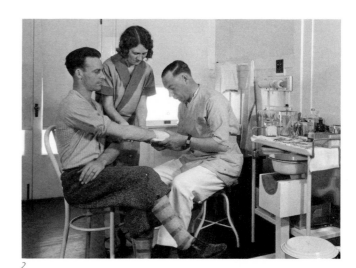

2

At its simplest, the optical printer is essentially a camera and projector mounted facing each other on a common axis. The basic function of the printer is to duplicate existing footage by rephotographing it as it is projected. As such, it is a basic tool in the production of special photographic effects and the range of camera tricks known collectively as "opticals" — fades, dissolves, wipes, freeze frames, etc.

For example, the optical printer operator can create a fade-out by running the scene and gradually reducing the amount of light available to the camera until it records only black; reverse this process and a fade-in is created. A freeze frame is achieved by stopping the projector at the chosen frame and photographing the still image over and over again with the camera. The optical printer can also be used to superimpose titles and credits over live action.

Modern optical printers are extremely sophisticated machines capable of combining as many as ten or more separately projected images onto one piece of film. With such capability the operator can create composite images, such as matte paintings with live action elements, or achieve split screen effects showing multiple actions.

The optical printer operator is usually employed by an independent laboratory or special effects company and may work under the supervision of the film editor or the special effects coordinator. (IATSE, NABET)

Playback Operator

The playback operator, working under the supervision of the production sound mixer, is responsible for the playing of any pre-recorded music or dialogue on the set.

During a concert scene, for example, singers or musicians fake their performances (lip-synching or pretending to play an instrument) in synchronization with the pre-recorded music being played over a sound system by the playback operator. During such a scene no sound is recorded on the set; the pre-recorded music will be edited to the image during post-production.

The playback operator sets up and operates the playback equipment and speakers, maintains proper volume levels, and must ensure that his equipment does not interfere with any other equipment or personnel. He works only when playback is required and is rarely on the set during the entire course of shooting. (IATSE, NABET)

Post-Production Supervisor

The job of post-production supervisor may be filled by a variety of people, including a production manager or film editor. The post-production supervisor acts on behalf of the producer, handling administrative details, coordinating activity between the director and editor, scheduling ADR sessions, and generally keeping the various post-production processes moving along on schedule and within budget.

Process Projectionist

Process photography refers to the two techniques known as front and rear projection. In both processes, an existing piece of film is projected onto a special screen erected on the set. Process photography is supervised by the director of photography, who often gathers the background "plates" – the footage to be projected – either from stock footage libraries or by shooting it to the film's requirements. The actors (and whatever sets or scenery are necessary) work in front of the screen: The live action and the projected image are then photographed as a unified whole by the camera.

In rear projection, the projector is placed behind a translucent screen on the same axis as the camera. Front projection uses a more complicated setup wherein the projector is placed at a ninety-degree angle to the camera; the projected image is reflected to the screen by a half-silvered mirror, which the camera is capable of shooting through. Perhaps the most familiar use of process photography is the shot, seen so many times, of actors in a car as street scenes glide by through the rear window.

On the set, the process projectionist works in a team with the director of photography and the rest of the camera crew. He is responsible for set up and maintenance of the screen and projector. He loads, unloads, and operates the projector at all times during its use.

Projectionists work in many places – theaters, laboratories, scoring stages, re-recording studios – but process photography is the only job that brings the projectionist to the set. (IATSE, NABET)

1 Bill Elliott and Walter Brennan ride the range without leaving the studio for a scene in Sleep All Winter, 1949.
2 Clark Gable, Spencer Tracy, and Ralph Morgan pretend they're on an oil field for a scene in Boom Town, 1940.

2

The producer is the person most responsible for the nuts and bolts of a production and shares with the executive producer the ultimate financial responsibility for the success or failure of a film or television production. Sometimes he will inherit an idea or property from the executive producer, but it is equally possible that he develops a concept or property himself. In either case, his involvement begins at or near a project's inception and continues throughout the life of the production.

With the property chosen, the producer hires a writer or writers to produce the final shooting script. In television this stage may take a wider variety of forms: A dramatic series or situation comedy may require the writing of a pilot episode; if a series becomes established, writing is a continuous process with possible scripts coming from both an in-house staff and freelance writers. At this early stage the producer will also hire a director, with whom he will work very closely

3 **Samuel Goldwyn** *stares down the camera for this publicity still with Ronald Colman and Vilma Banky, c. 1927.*

4 **David O. Selznick** *(center) makes a last minute script check with Director John Cromwell and Script Supervisor Cora Palmatier for* Since You Went Away, *1944.*

4

3

throughout the production. When shooting begins, the producer usually steps aside, leaving the bulk of the authority in the director's hands while still closely monitoring the schedule and budget.

Television is often called a "producer's medium" because the producer or producers may contribute more to the style and content of a show than any individual director. A series, for example, may employ a number of directors on the basis of individual shows or in rotation. The bulk of a producer's time during planning and production is spent with administrative responsibilities. After overseeing the writing and selecting a director, he hires other crew members, supervises casting and negotiates actor salaries, finalizes the budget and the schedule, and hires facilities, equipment, or studio space as needed. If a crew goes on location, the producer must make all travel and lodging arrangements and secure permission from individuals or city governments.

The producer also monitors post-production work and, for a feature film, may be involved in plans for distribution and marketing, while a television show will require interaction with the network.

The producer is usually assisted by several members of a support staff. A team of two

1

2

1 **Irving Thalberg** *poses with MGM studio executive Louis B. Mayer, c.1925.*

2 **Daryl Zanuck** *strikes a thoughtful pose for this publicity still from* The Longest Day, *1962.*

3 **Hal Wallis** *(right) poses with studio executive Jack Warner and Director Michael Curtiz (standing) around the script of* Casablanca, *1942.*

4 **Arthur Freed** *and Maurice Chevalier review a costume sketch for* Gigi, *1959.*

5 **Dino De Laurentiis** *surveys an enormous Roman set for* Barabbas, *1962.*

5

3

4

producers is not unusual, and there is often a large production staff comprised of associate producers, assistant producers, or line producers. Their function is to assist the producer in carrying out his responsibilities, and the scope of their authority and the division of duties among them varies widely.

The producer's creative contribution may be very great or very small. The producer who takes an active part in the supervision of casting, writing, design, and editing may exert a considerable influence on the style and content of the finished production; other producers may concentrate on administrative and financial responsibilities and leave the creative decisions to others.

Fred Coe *(right) and Director Arthur Penn confer on the set of* The Miracle Worker, *1962.*

. . . .

Production Assistant

Production assistant is a catchall term and the responsibilities involved depend greatly on the individual production. The production assistant might spend most of his time running errands or he may have a position of responsibility with the producer, director or assistant director, or production manager. Working as a production assistant is an ideal way for people interested in film or television to get their first experience.

Production Associate

The production associate, working in taped television production, holds a position somewhat analogous to that of the script supervisor in film production. When the production associate reviews the script of an upcoming program, he or she creates brief production breakdowns for all departments: The script is divided by scene, performers, time of day, costumes, props, and notes to the sound and camera crews. The production associate also keeps the script up-to-date and distributes script changes and rewrites as they occur.

Another important aspect of the production associate's job is timing: Television programs usually must fill precise time slots. He or she makes a time estimate from the rough copy of the script and, during later rehearsals and taping, records the actual time of each scene and informs the director if there is too much or too little material.

The production associate keeps an eye out for continuity, alerting the crew to any inconsistencies or mistakes. The production associate works primarily for the director and is in frequent contact with other key personnel — the technical director, associate director, stage manager — by headset. (DGA)

Production Auditor/Accountant

Even a relatively small film or television project generates an enormous amount of administrative responsibility, and it is the production auditor who manages the day-to-day details of a production's financial resources.

When the producer has a script and a director and pre-production planning is set to begin, the production auditor is engaged. Although ultimately responsible to the producer, he works closely with the production manager. He must know the script and the budget and be prepared to help the production manager revise the budget, as well as provide expense reports and estimates as needed. He opens bank accounts, sets up an accounting system for the production company, pays bills, pays salaries to the staff, crew, and cast, handles insurance claims, and, at the end of production, closes all books and accounts and distributes all records to the designated party. (IATSE)

Production Designer

The production designer has the ultimate artistic responsibility for the design of a film project. He must translate the director's ideas into an actual physical environment. Depending on the film, this may involve constructing sets, finding the right location and decorating it to suit the production, or both. A key member of a film's creative team, the production designer begins work early in pre-production. He reads the script in terms of set and location requirements, and arrives at a budget estimate that he must negotiate with the producer (who has made his own budget estimate). Because the designer administers a large staff and a potentially enormous budget, he owes a strong fiscal responsibility to the producer, but his creative responsibility is solely to the director.

With a budget, a complete breakdown of set requirements, and continuing creative input from the director, he and his staff prepare drawings and plans, build three-dimensional models, and scout locations, making changes and refinements until the director and producer grant their approval.

The designer provides the working drawings from which the sets are built and supervises construction, property acquisition, and set decoration throughout shooting. The

2

1

designer is present on all locations, where he may oversee a set as simple as artificial greenery or as complex as a new building.

The designer must closely coordinate his work with the director of photography and the costume designer. In designing the sets he must take into account the cinematographer's plans for lighting and camera movement and, with the costume designer, he works to achieve a visual integration of color and pattern between sets and costumes. (IATSE, NABET, USA)

1 Designing sets for the "Eddie Cantor Show" at NBC, c.1950.

2 **Lyle Wheeler** (right) and Producer Frank Ross inspect a set model for The Robe, 1953.

3 **Van Nest Polglase** (second from right) with staff at RKO.

4 **Tony Walton** (left) conferring over a set model with Producer Rob Cohen (center) and Director of Photography Ossie Morris for The Wiz, 1978.

4

3

Production Manager

The production manager, among the first people hired by the producer, assumes the enormous responsibility for the day-to-day planning and management of the business side of film or television production.

The production manager's first job is to figure out a budget, based on the requirements of each department – sets, costumes, props, equipment, cast, transportation, etc. The production manager may also work out a preliminary shooting schedule, and he will work very closely with the assistant director during the formulation of the final shooting schedule. He hires the technical crews and works out their terms of employment, and makes deals and draws up contracts or agreements with the vendors providing equipment, supplies, and services. He has considerable involvement in the selection of locations and will handle arrangements for permissions and releases, scouting, travel and transportation, and catering, or turn them over to a location manager.

During pre-production and shooting the production manager monitors all aspects of the production and tries to keep it on schedule and on budget. He prepares production reports documenting the work completed each day.

The production manager is directly respon-sible to the producer and director, who rely on him to run the production smoothly and steer it through the problems and logistical pitfalls that can waste precious time and money. To this end the production manager also works closely with each department head, helping to assemble resources in equipment, personnel, and supplies.

The production manager must have an extensive knowledge of all facets of production – from the activities of each department to union regulations and pay scales. He usually remains through at least part of post-production to settle business affairs, and sometimes becomes the post-production supervisor. (DGA)

1 **Keith Weeks** makes arrangements for the Pismo Beach location of Strange Cargo, 1939.

The production office coordinator serves as a "clearing house" for all the paper records of a production and sees that they are prepared, maintained, and distributed in the proper form and to the proper places. This involves a wealth of detail — for example, preparing and distributing production breakdowns, scripts, and script revisions, shooting schedules and call sheets; maintaining the weekly or daily production and financial reports; and seeing that camera reports accompany the negative to the lab and continuity sheets get to the editor.

The coordinator must have a working knowledge of all contracts and agreements for cast, crew, writers, and equipment and service suppliers, and may prepare and execute them as required by the producer or production manager. Pre-production duties may include filing permits and making travel and eating arrangements for cast and crew.

On feature films, the coordinator may work well into post-production to wrap up production office business, dismantle the office, and see that equipment is returned and records distributed as desired by the producing organization. (IATSE)

2 **Adeline Leonard**, *Director Sydney Pollack, and Producer Stanley Schneider on location for Three Days of the Condor, 1975.*

2

The production sound mixer is responsible for all sound recorded live on the set or on location. Ideally, the sound mixer consults with the director (and possibly the supervising sound editor) and checks all locations well before the start of shooting to determine his goals and the equipment needed.

During shooting, he sets up the equipment, determines the placement of microphones (including small wireless microphones concealed on the actors), monitors sound, and operates the recording equipment. He takes part in all rehearsals and must coordinate with the camera and electrical crews so that camera, lighting, and sound equipment do not interfere with one another. The sound mixer often works under difficult conditions and must alert the director of any extraneous noise or other problems that interfere with effective recording. (IATSE, NABET)

1 Paul Newman being wired for sound on the set of Slap Shot, 1977. Director George Roy Hill sits at left.

2 **H. Townsend** in front of the soundproof recording booth used in Nothing But the Truth, 1929, one of the earliest sound films made at the Astoria Studio. Inside the booth are (left to right) Camera Operator Ed Cronjager, Sound Technician Ernest Zatorsky, and Director Victor Schertzinger.

3 On location for Our Daily Bread, 1934. Director King Vidor sits atop the ladder.

3

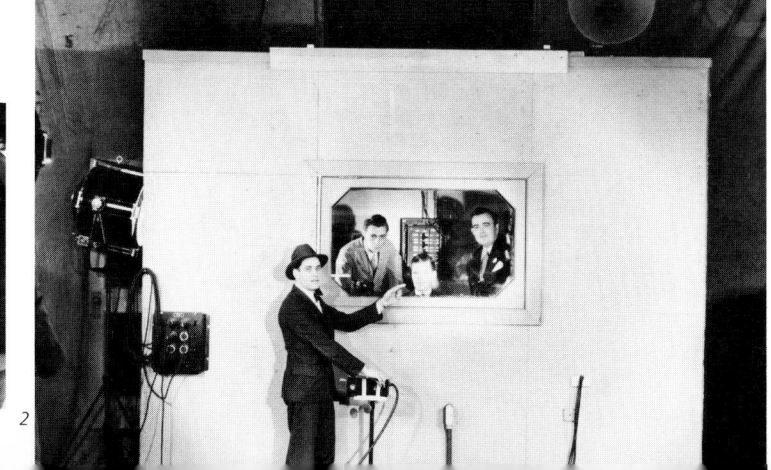

1

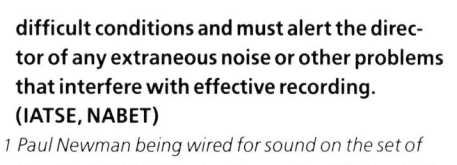

2

The projectionist's job is the final step in a film's journey from conception to screen, and it is as important as any other. All of the time, money, and effort invested in a film can be severely compromised if a film is poorly projected. To get the best possible image on screen, the projectionist must attend to both film and equipment: When a film arrives, the projectionist inspects it, repairs any damage, and then cues it up to its starting point, making sure it is projected in frame, in focus, in sync (sound and image), and at the proper volume.

Feature-length films are made up of several individual reels and projection booths are equipped with at least two projectors to allow smooth changeovers from one reel to the next. A warning bell followed by a mark on the film signal the end of a reel and tell the projectionist when to roll the next. While the second reel runs, the projectionist rewinds and removes the first to make room for the next, repeating this process until the last reel is shown.

Theaters today often have sophisticated new equipment that performs changeovers automatically or semi-automatically. This equipment allows a projectionist to run as many as three films simultaneously, a vital necessity in the age of multiplex cinemas.

Projectionists work in a number of places besides theaters. As projection is required in such processes as re-recording, ADR, music recording, and laboratory inspections of newly printed film, projectionists are employed whenever and wherever these processes take place. (IATSE, NABET)

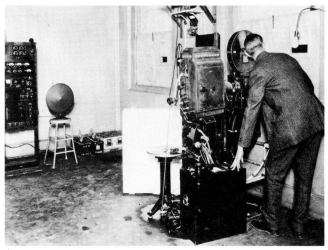

4 This much-touted projector could synchronize recorded music with film a few years before talkies, 1926.

Property Maker

Although most props can be bought, rented, or borrowed, some must be specially made to suit the needs of a particular production. A special prop can be nearly anything, from sculptures meant to be the work of an artist character to a tombstone made of styrofoam, to "breakaway" items (balsa wood chairs and tables, candy glass bottles) for use by actors and stunt performers.

The work of the prop maker may be assumed by a number of different people. The property master is usually capable of building some special props. Others will be built by special effects experts. The property maker works closely with the director and production designer throughout design and construction. The prop maker may be responsible to the property master, the production designer, or the director. (IATSE, NABET, USA)

1 Making pottery at MGM.

1

The property master's primary responsibility is to acquire, maintain, and oversee the use of all the hand props indicated in the script. Hand props, in general, are those items actually used or handled by the actors during a scene. For instance, if a living room bookcase is filled with books that will never be touched by the actors, the shelves will be stocked by the set decorator; but if the script calls for a character to pluck out a copy of War and Peace and hurl it at another character, it becomes the responsibility of the property master.

Depending on the production, the property master might provide anything from old telephone directories to statues, paintings, cars, guns, or more exotic weapons – anything the actors will use in the course of filming or taping. Props are culled from a variety of sources – prop rental houses, department stores, flea markets, antique stores – and experienced prop masters often develop their own directories listing useful suppliers.

During pre-production, the property master reads the script, prepares a rough breakdown of the required props, and often

2 **Robert Martin** whips up instant cobwebs for a scene in One Foot in Heaven, 1941.

3 **Stanley Dunn** with his decorated prop box at Columbia Pictures, 1953.

2

3

provides the producer with a budget estimate. Much of his planning must be coordinated with other departments and, if stunt work is involved, he will work with the stunt coordinator to determine requirements for stunt cars, special "breakaway" props (balsa wood furniture or candy glass bottles), and trick knives or other weapons.

During shooting the property master and his staff supervise prop placement on the set, maintain and repair props as needed, and often keep detailed continuity records to complement those of the script supervisor.

The prop master also handles some atmospheric effects – smoke, fog, wind – but calls in a special effects team for large scale effects.

The prop master is responsible to the director and sometimes to the production designer or art director. (IATSE, NABET)

1 In the prop room at Columbia Pictures with assorted guns and Gene Autrey's guitar.

· ·

Assistant Property Master

One or more assistant property masters help carry out the property master's duties during shooting. The assistant's responsibilities usually include the pick-up and return of props and their positioning on the set. The assistant also inventories and maintains the prop room, performing minor repairs and cleaning as necessary. Often, one assistant is in charge of continuity – writing notes and taking Polaroids to document the appearance and placement of props during shooting. (IATSE, NABET)

1

Prosthetic Makeup Artist

The producer hires a prosthetic makeup artist when a production involves extensive special makeup – elaborate character transformations, the creation of monsters and mutants, or the preparation of close-up and gory wound and blood effects. The prosthetic specialist and the director meet in pre-production to develop the director's ideas into workable plans for the makeup. Most prosthetic specialists are skilled sculptors, and much of their work involves the creation of new forms and surfaces.

The prosthetic artist begins by making a life-cast of the actor to work with while designing the prosthetic appliances. When the designs are approved, the specialist molds and constructs the makeup appliances out of soft foam latex, plastics, cloth, hairpieces, metals, or other materials. For wound effects he may use "blood" sacks and tubes concealed under a layer of rubber "skin." He also paints the pieces in the appropriate human or other-worldly colors.

During shooting, he is usually on the set only when the special effects are used, but on those days he and the actors concerned are probably the first to arrive and the last to leave: Prosthetic makeup can take over five hours to apply and three to remove. In addition to working with the director and actors, the prosthetics specialist may also work with the costume designer, head makeup artist, hairstylist, and very often with other special effects experts, especially with makeup for wound and injury effects. (IATSE, NABET)

2 **Tom Burman** *touches up a nightmare creature for* Cat People, *1982.*
3 **Rick Baker** *makes up Griffin Dunne for* An American Werewolf in London, *1981.*

3

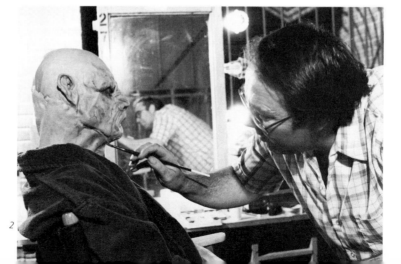

2

Pyrotechnic Specialist

The pyrotechnic specialist is a special effects expert highly skilled in creating real or illusionistic explosive and fire effects — bombs, exploding cars, burning buildings, simulated bullet hits on actors, walls, or furniture, or even burning stunt performers fleeing from disasters.

After reading the script and consulting with the director, the pyrotechnic specialist prepares a rough breakdown and estimates a budget. The planning phase can be extremely complex and may require extensive cooperation with other departments. In a war movie, for example, where a battlefield will be pockmarked by gunfire and bombs, the pyrotechnic expert will conceal charges underground, on rocks, or buildings, etc., and plan a sophisticated network of wires leading back to a control station where the charges will be detonated. This may well involve mapping out the location of each charge and extensive cooperation with the assistant director, who must know where the charges are so that he can get crowds of soldier-extras through the exploding battlefield without harm.

Similarly, bullet hits on actors may involve coordination with the wardrobe and makeup departments, as well as the actors themselves. Bullet hits are often small explosive charges mounted on metal plates (to protect the actor) concealed under the actor's clothing. If the bullet is to hit an area of exposed skin, the pyrotechnic expert will work with the makeup or prosthetic makeup artist to plan a device that will conceal the explosive charge and the wires while still retaining a natural appearance.

Fire effects, such as a burning building, are usually carefully controlled illusions; though it has been done, buildings or other structures are rarely blown up or burned for real. Instead, the pyrotechnic expert will use a system of gas jets (usually propane) that can be turned on and off. In such cases he may need to work with the production designer, whose plans for the doomed set will have to accommodate the pyrotechnic apparatus.

The pyrotechnic expert is licensed by state and federal governments to handle and use explosives, and he must also act as the safety officer when the effects are performed. In addition to providing his own fire-fighting equipment, he often must arrange to have emergency personnel on the set or location. (IATSE, NABET)

1,2 On location for The Traveling Executioner, 1970. Setting explosive charges and standing by the detonator box as the inferno begins.

Recording Engineer

The recording engineer works in the recording studio — as opposed to on the set or location — during recording sessions. He may work under the supervision of a wide variety of people — the director, producer, or the film, music, sound, or ADR editor — but he is usually employed by an independent recording studio hired by the production company. The recording engineer is responsible for setting microphones, maintaining the recording equipment, and for operating the equipment so as to achieve the tonal qualities desired by the filmmakers. (IATSE, NABET)

· ·

Re-Recording Mixer

When the film is finished and edited, there may be fifty or more individual sound tracks — dialogue, sound effects, music tracks — and the re-recording mixer's job is to balance them relative to one another, manipulate them, and combine them by re-recording onto a single sound track.

Re-recording is performed in a specially equipped studio where, seated at a control panel, the mixer can adjust each track independently of the others as the film is being screened. The mixer works from cue sheets in the form of charts, which tell the mixer where each sound effect or musical passage begins and ends and offer instructions and suggestions on volume levels, fades, etc.

The mixer can also manipulate sounds. For example, he can make a voice sound as if it

3 Mixing the sound track of The Catered Affair, *1956.*

3

is coming over a telephone or from a distant room. The mixer usually has plenty of company during the re-recording sessions: The director, producer, film editor, sound editor and music editor, plus assistants, may be present for all or part of the process to provide creative input. Adjustments between the sound tracks are refined gradually and usually practiced many times before the right combination is achieved.

Re-recording is generally a very slow process and, on a feature film, the mixer might finish just five minutes, or less, of screen time per day. The mixer is ultimately responsible to the director or producer, but, in their absence, usually answers to the film editor. (IATSE, NABET)

1 Mixing the sound track of <u>Apocalypse Now,</u> *1979.*

1

The scenic artist is an important member of the team of artisans and craftspeople who execute the plans of the production designer. Working under the supervision of the art director, he or she may be responsible for the application or painting of all decorative wall or surface coverings, portraits or any other special artwork, lettering and sign work, backdrops, or the aging of surfaces, sets, props, and even costumes. Since the scenic artist's work is often dependent on the completion of the set and the acquisition of props, he or she must closely coordinate scenic work with the construction and property departments.

On a large production with many scenic artists, a chargeman scenic artist supervises this staff and maintains the closest working arrangement with the art director or production designer. At least one scenic artist remains on the set throughout shooting. (IATSE, NABET, USA)

2 Touching up a backdrop for _Wolfen_, 1981.
3 Preparing an eighteenth-century French wall panel for a set in _Scaramouche_, 1952.

2

3

Script Supervisor

The script supervisor keeps a detailed log of each day's shooting activity. Stationed near the camera, he or she records the action of the performers, their positions and attire, the dialogue spoken, the number and duration of takes, the nature of camera movements, the lenses and filters used, and tells the second assistant camera operator the number of each take.

On the set the script supervisor is directly responsible to the director and the director of photography, but the continuity sheets, the highly organized records he or she produces, serve as important guides to the assistant director, the costume designer, the property department, and eventually the film editor. (IATSE, NABET)

1 *The World of Henry Orient*, 1964.
2 *Man on a Swing*, 1974.

1

2

The second assistant camera operator carries out a variety of functions during shooting. Acting under the instructions of the director of photography, his primary duties are loading the film magazines, operating the clapper board, and maintaining a paper record of all shots and takes.

The second assistant loads the film in total darkness to avoid exposure and makes sure that a fully prepared magazine is on hand to replace the one in use. He keeps track of the amount of film in the camera, and alerts the crew when there is insufficient film to cover the next shot. At the beginning of each shot he calls out the take number and operates the familiar clapper board (hence the more familiar title of clapper/loader), which he must also keep up-to-date. This procedure serves as a visual and aural identification of each take.

One of his most important duties is a paper log – called "camera sheets" – listing all the scenes and takes completed on each day of filming. These sheets are an important reference for both the laboratory and the film editor, who need to keep track of enormous quantities of footage. The second assistant position requires some prior film experience and a technical knowledge of the camera equipment; it may lead to advancement through the ranks of the camera crew. (IATSE, NABET)

3 Getting set for a scene with Joan Fontaine in *Frenchman's Creek*, 1944.

3

Second Assistant Director

The second assistant director is the right hand of the first assistant director during pre-production and shooting. He helps carry out the first assistant's responsibilities and, depending on the authority delegated to him, may assist with location scouting and selection, extra casting and hiring, direction of background action, notifying cast and crew of scheduling, and making sure everyone is on the set and in place on time. The second assistant is generally responsible for preparing and distributing the daily call sheets, which he prepares from the master schedule.

The second assistant may also have assistants (called second second assistant directors), and supervises the DGA trainee, if any, on the production. (DGA)

Second Unit Director

Though some directors insist on shooting every shot in a film themselves, many find it expedient or necessary to use a second unit director in certain situations: atmospheric shots not involving the actors, such as the sunrise over an African landscape or a city street scene at rush hour, or large-scale action scenes, a particular specialty of some second unit directors.

During pre-production, the second unit director meets with the director, producer, director of photography, assistant director, and production manager to plan the particulars of the scenes he will shoot, develop a schedule, and determine and arrange for equipment and crew requirements. He has full directorial authority when out working with the second unit but his output must serve the director's overall vision of the film and match the color values and lighting style of the work of the main unit.

Depending on the project's complexity, the second unit director may work with a small camera crew, or a miniature version of the main unit crew, with a second unit director of photography, camera crew, gaffer, electricians, and sound crew. (DGA)

Set Decorator

The set decorator, under the supervision of the production designer or art director, coordinates the furnishing of the set. On a modern set requiring relatively simple, contemporary furnishings, the decorator might not begin work until the sets are ready to be built. On a production depicting a distant historical period, however, the decorator may be hired early in pre-production to allow time for extensive research and the acquisition of more obscure decorative items.

Using the shooting script to determine what furnishings the production requires, the decorator conducts research, acquires the items, and works out their placement on the set. Since responsibilities sometimes overlap, the set decorator must coordinate his or her work with the property master; he or she must also work with the director of photography, whose plans for camera placement and movement will directly affect the arrangement of set furnishings.

During shooting, the set decorator is often away from the set, searching for and acquiring furnishings one step ahead of the shooting schedule. In his or her absence, placement of the items on the set is handled by the set dresser. (IATSE, NABET)

Set Dresser

The set dresser, under the supervision of the production designer and art director, is responsible for the placement of furniture and other decorations on the set. He or she also works closely with the property department; the responsibilities of the two departments often overlap and always require integration. The set dresser maintains the items, performs minor repairs as necessary, and relies on the continuity sheets of the script supervisor to ensure that set decoration is consistent from scene to scene. (IATSE, NABET)

Sideline Musician

Sideline musicians work on the set during shooting but they don't actually play their instruments. If, for example, a production calls for a nightclub scene with a small orchestra, sideline musicians fake their playing to pre-recorded music, which is edited to the image during post-production. **(AFM)**

1 *The orchestra assembles for a nightclub scene in* Alexander's Ragtime Band, *1938. Alice Faye sings, Tyrone Power conducts, and Don Ameche plays the piano. Musical Director Alfred Newman can be seen at right rear.*

1

Special Effects Coordinator (Visual Effects Supervisor)

Certain films might be classed as "special effects movies." They are fairly easy to spot, often set in distant galaxies. From Star Wars to Spaceballs to Ghostbusters, these films so depend on the skillful execution of extraordinarily complex effects that an entire effects crew — specialists in miniatures, mechanical effects, pyrotechnics, matte painting, animation, model making, and special photography — in addition to the main unit, is hired by the filmmakers to handle them. This crew is usually headed by a special effects coordinator.

During pre-production the coordinator works with the producer and director to plan the effects and determine a budget, and will continue to work closely with the director and other members of the main unit — director of photography, production designer, etc. — throughout production.

The coordinator's job is to marshal all this talent into an efficient unit for accomplishing work that always presents new challenges and problems: No two major special effects jobs are ever the same. Though the coordinator has a great degree of independence and creative authority (and sometimes runs his own company), the output of the effects unit must combine seamlessly with footage shot by the main unit. The coordinator often has considerable camera experience, and many coordinators have invented new equipment and techniques in the course of their work; for example, the computer-regulated motion control camera was invented by John Dykstra for Star Wars.

The coordinator may oversee post-production procedures with the effects footage — for example, editing and optical work — before it is turned over to the first unit editor. The special effects coordinator needs to be part technician, part craftsman, part engineer, and part artist, and the field is dominated by a few acknowledged masters such as Richard Edlund and John Dykstra. (IATSE)

2 **John Dykstra** with the optical printer used for the ABC television series "Battlestar Galactica," 1979.

Stagehand

On taped television productions, the stagehand performs duties comparable to those of the grip in film production. The stagehand puts together and positions sets and backdrops, pushes camera dollies or microphone booms, and may assist the lighting director in positioning lighting equipment and power cables. The stagehand is responsible to the stage manager, the lighting director, or property master. (IATSE, NABET)

. . . .

Stage Manager

In taped television production, the stage manager is the studio-floor representative of the director. He acts as a liaison between the director in the control room and the crew on the set. He works with the art director and property master to ensure that sets are prepared and also ensures that performers make it through the hair, makeup, and costume departments in time for their scheduled appearances on the set.

During rehearsals and taping, the stage manager is in constant communication with the control room by headset. He must have a thorough knowledge of the script and the director's staging plans since he is often called upon to cue the camera operators and the performers, which he does silently during taping using hand signals. He may also relay directions from the technical director to crew members on the studio floor.

The stage manager is one of the most important members of the television production team — a vital link between the control room and the stage. (DGA, IBEW, NABET)

Stand-ins substitute for principal actors while the camera and electrical crews go through the lengthy process of setting up and adjusting equipment for individual shots. Stand-ins are selected for their physical resemblance to the actors they replace. They are fitted with identical costumes and, on the set, assume the positions to be taken by the actors when the scene is shot.

By using stand-ins, the production can be sure that the principal actors stay fresh and unwilted for the actual shooting. Stand-ins sometimes serve as on-camera substitutes in extreme long shots, but for stunts or potentially dangerous situations, a stunt double is hired. (AFTRA, SAG, SEG)

1 **Bob Keef** and James Cagney pose on the set of A Midsummer Night's Dream, 1935.

2 **Sally Sage** and Bette Davis pose on the set of The Petrified Forest, 1936.

1

2

The Steadicam operator is a skilled camera operator who specializes in the use of the Steadicam system. Steadicam is not a camera, but rather the system that supports the camera, and nearly any lightweight camera can be attached to it. In the system, the operator wears a vest or body harness that redistributes the camera's weight to the operator's hips. The camera is attached to the harness by means of a spring-loaded arm that damps and smooths the motion of the camera, providing an image steadiness comparable to a dolly or tracking shot. Further, the image recorded by the camera is relayed to a small video monitor attached to the system, thereby freeing the operator from the camera's eyepiece.

The advantage of the Steadicam system is the range of mobility and flexibility it affords without compromising image quality. Providing it fits, it can go nearly anywhere the operator's two feet can take him, for example, up and down stairs, through subway cars and back alleys, and down narrow corridors.

The Steadicam operator will consult with the director and director of photography during pre-production and continue to work with them as needed during shooting. As must any camera operator, he works in close cooperation with the camera, sound, and electrical crews. (IATSE)

1 Shooting a nighttime action sequence for Bad Boys, 1983.

1

Actor portraits, photographs that record the action of a production, and behind-the-scenes shots of the crew are used in the press kits sent to newspapers, in magazine articles, in theater lobbies and windows, advertisements, posters, on-air promotions, and any other place where they can be used to generate public interest.

On feature productions, the still photographer is present on sets and locations for the duration of the shooting phase. He or she may also perform limited pre- and post-production work, for example, photographing wardrobe and makeup tests. Work for television shows usually involves much shorter assignments as determined by the producers or network.

The still photographer also takes any photographs that are to be used in the production, such as snapshots of the characters that appear on bedside tables or the photos of the bad guys that have passed over many a screen detective's desk.

For any shots taken off the set, such as actor portraits, he or she determines lighting setups. Lighting on the set is the province of the director of photography, and the still photographer must not interfere with the progress of shooting. (IATSE, NABET)

2 **Muky Munkacsi** on the set of Child's Play, 1957.

2

Storyboard Artist (Production Illustrator/Sketch Artist)

When the production designer has roughly mapped out the main sequences of action and has prepared preliminary plans of the major sets, a storyboard artist may be called in. Working closely with the director and production designer, he translates their ideas into storyboards – rough sketches of scenes from the film in sequential order. The sketches, each depicting a key narrative element, can be arranged to approximate a sequence of shots in the film, and thus are a useful tool for planning and troubleshooting before construction and shooting begin.

Storyboards are widely used in animation, complex special effects sequences, and the preparation of television commercials; in most feature films and television their use depends on the preference of the director and the complexity of the project. (IATSE, NABET, USA)

Preparing a storyboard sketch.

Stunt Coordinator

Film and television productions involving many stunts rely on a stunt coordinator to plan, organize, and supervise all stunt work and stunt performers. Brought on during pre-production, the coordinator may have considerable creative input during the planning phase. Though some scripts are very detailed and some directors very sure of what they want, other scripts might simply specify "chase scene," and the director may ask the stunt coordinator to design it.

The coordinator chooses stunt players and doubles, and often performs stunts as well. He devises a budget by breaking the stunts into specific needs, and then negotiates that budget with the producer or production manager, who inevitably wants to spend less. The coordinator will also have to communicate his plans to department heads: If there will be a brawl in the mud, the wardrobe department must prepare multiple wardrobes for retakes; if there's a chase scene and subsequent highway pile-up, the property department must know how many cars or other vehicles will be needed. The stunt coordinator also works closely with special effects experts, who rig up many of the catastrophes befalling the stunt players.

One of the stunt coordinator's greatest concerns is safety. His primary and best defense against accidents is the meticulous planning of each stunt with an eye to every conceivable problem that could arise. Although medical personnel stand by in the event of the unforeseen, the stunt player's livelihood depends on his ability to dust off and do it again the next week. (AFTRA, SAG)

Stunt Person

The most dramatic expression of the stunt person's job appears on theater and television screens every day. Stunt people are performers and, though they are often the anonymous victims of any number of catastrophes, they may also have small speaking parts. A stunt person serving as a stand-in must share a physical resemblance – height, weight, build, etc. – with the actor he or she replaces, and is fitted with an identical wardrobe.

Stunt people take great pains to ensure they don't meet the gruesome or messy fates they simulate for the camera. Stunts are carefully planned and accompanied by extensive safety precautions, including ambulances and fire-fighting equipment. Stunt people work closely with the director and the special effects experts, who rig up many of the traps they fall into. On productions full of stunts, stunt people work for the director and a stunt coordinator. (AFTRA, SAG)

1 *Biting the dust on location for* The Scalphunters, *1968.*

1

The supervising sound editor's principal responsibility is the creation of the film's sound track – the combination of dialogue and sound effects tracks into an effective aural complement to the film images.

Ideally, the sound editor becomes involved on a film project during pre-production when, based on the script, he can offer suggestions to the sound mixer concerning what extra sounds to record during shooting. His work really begins, however, when the film editor completes a rough cut of the film. At that point the sound editor may put together a temporary sound track by adding existing sound effects and music to the dialogue. This "temp track" allows the director to evaluate the film to get a better idea of the effects he wants the sound editor to achieve. This version of the film is sometimes shown in previews to test audience response.

The sound editor's first priority is the dialogue, which often amounts to half the job. Dialogue recorded during filming is usually incomplete; extraneous location noise often renders portions unusable, and some location shooting makes recording impossible. The sound editor hires an ADR editor to record missing dialogue and begins to assemble the sound effects tracks – everything from footsteps to squealing tires to machine gun fire. For this he has four major resources: sound effects libraries, the tracks recorded during filming, those he records himself, and those recorded by the Foley artist. The last requirement is the music track, which the composer/conductor and music editor complete independently.

When all tracks are complete, the sound editor prepares the cue sheets, which tell the re-recording mixers where each sound effect begins and ends and offer instructions about such things as fading and volume. Though primarily the province of the re-recording mixers, the sound editor usually remains involved throughout re-recording to offer ideas and suggestions. (IATSE, NABET)

2 In the editing room with library sound effects.

2

Apprentice Sound Editor

The apprentice sound editor is responsible to the assistant sound editor or sound editor. He or she helps with office chores, runs errands, and perhaps helps with filing, cataloguing, and keeping records on the film and sound tracks in an effort to learn and observe as much as possible. (IATSE, NABET)

. . . .

Assistant Sound Editor

Film productions may require the skills of several sound editors working under the supervising sound editor. Each sound editor is assigned to a particular reel of the film and each has an assistant sound editor.

The assistant provides a variety of support functions, from keeping the editing room in order, to maintaining extensive records, to synchronizing film and sound tracks. Sometimes the assistant acts as the sound editor's representative at Foley or ADR sessions. Assistant sound editors usually accompany the sound editor to meetings, re-recording sessions, etc., bringing all necessary material and taking notes as needed.

The assistant's position is generally a stepping stone to a position as sound editor and, under close supervision, he usually performs some editing as well. (IATSE, NABET)

The technical director is ultimately responsible for ensuring the technical quality of taped television productions. He oversees the setup and adjustment of control-room equipment, video cameras, videotape recorders, audio equipment, etc., and supervises the personnel who operate them. The technical director works in the control room during rehearsals and taping, and is in communication with other key studio personnel (director, stage manager, camera operators) by headset.

The technical director's other responsibility is the operation of the switcher, the control-room unit that determines which camera images and special effects (such as fades and dissolves) are sent out over the air or recorded. The technical director operates the switcher according to instructions from the director, who is usually next to him in the booth. Switching involves communication with the camera or video operators to ensure smooth and accurate transitions from one shot to the next. On news programs, the technical director must supervise the smooth integration of taped or live remote material into the broadcast. A substantial amount of technical know-how and several years experience are prerequisites for the position. (IATSE, IBEW, NABET)

In the control room during the first telecasts of the House of Representatives, 1979.

115

Theater Owner/Manager

The theater owner/manager selects films for exhibition, publicizes them, and supervises the daily operation of the theater. Some theater owners hire a film booker to choose films; others select their own by reading the trade press, attending advance screenings, and making deals with distributors. Publicizing a film involves obtaining materials — lobby cards, posters, newspaper advertisements — from the distributor and placing them in the theater lobby and local newspapers.

The theater owner also hires and supervises theater employees such as projectionists, ushers, concession counter workers, and box office cashiers. He manages the theater's accounts, including paying the distributor his percentage of the box office proceeds.

The managers of theater chains differ in that the chain's central office handles bookings and acquires publicity materials.

Sid Grauman *watches Harold Lloyd cement his fame in the forecourt of Grauman's Chinese Theatre, 1927.*

Theme Song Composer

Nearly every television show has original music of some sort. Series themes, which sometimes make it to the pop charts, are the most elaborate, but even news programs have intros during the opening and closing titles. When a show is being developed for broadcast (or a new theme is needed for an existing show), the producer hires a composer to develop appropriate theme music and, perhaps, lyrics.

Theme music has its own genres and is conceived to convey the mood and tone of the show or even tell a little about the characters, as did the storytelling theme music for "Gilligan's Island" and "The Beverly Hillbillies." In such cases, the composer familiarizes himself with the characters and storylines in order to create a theme that meets the producer's (and perhaps the network's) approval. The composer may be involved with the recording and mixing of the theme music as well.

The term "titles" refers mainly to the opening and closing credits of film and television programs. Early titles were relatively simple, usually involving title cards designed to reflect the theme or setting of the production. Modern titles have become considerably more complex, combining graphic design with a variety of live action elements and sometimes animation.

The title designer usually works most closely with some combination of director, producer, and editor, and the first step is always to develop a concept. At this stage, he or she also works closely with the composer to integrate the music and the visuals. Though the visuals usually come first, in some cases the title designer must mold the visual sequence to the pace and rhythm of an existing piece of music. Though the director may shoot live action credit sequences with the main unit, it is equally possible that the title designer will assemble a crew and direct and edit these sequences him or herself.

When the scenes are shot and edited and the graphics designed, the title designer will usually contract an independent title company to combine the graphic and live action elements on the optical printer. These title companies also handle the entire job for less ambitious title sequences, for example, if the production only requires that lettering be superimposed over opening scenes.

The degree of the designer's creative contribution varies widely depending on the production: Sometimes the director or producer has already developed a concept; in other situations, the designer is asked to deliver a creative solution. (IATSE)

Transportation Coordinator

The transportation coordinator supervises the transportation of equipment, sets and scenery, and personnel for studio and location work in film and television. He is responsible to the production manager and meets with him during pre-production to determine the production's transportation needs and to plan for the proper number of vehicles and drivers. The driver's responsibilities range from local trips to and from the scenic warehouse, to cross-country treks with equipment, to chauffeuring directors and actors. (IBT, NABET)

. . . .

Treatment Writer

Film and television scripts may go through many hands and many lives before they reach either the big or little screen. The producer hires a writer to develop a treatment as one of the first stages in the birth of a production. Written in prose form, it describes the main action and might also include important passages of dialogue. The treatment writer may go on to write the final script, but many treatments never become scripts at all. (WGA)

The unit publicist acts as a liaison between the studio's or distributor's publicity department and the production organization making the film. The publicist usually starts working some time prior to shooting and sets about the job of getting the word out. He writes press releases and notes, short bios of the actors and director, and places stories and articles in newspapers and magazines both locally and nationally. He may arrange interviews with actors or directors, either on television or with reporters visiting the set.

The unit publicist works with the still photographer to plan publicity stills and sometimes with outside photographers sent by a magazine. He writes the materials for the press kit — bios of actors and key crew members, a synopsis of the plot, an account of how the production came to be made, and captions to accompany the photos.

No two publicity campaigns are alike, and the unit publicist, along with the producer and the studio, tailors the campaign to the particular project. On a production with major stars the publicity may be minimal until the project's release. On a production featuring unknowns, a more intensive campaign may be waged. The unit publicist must be careful to avoid overexposure: Release can follow the completion of shooting by a year or more, thereby diluting the effectiveness of an early publicity campaign. (IATSE, NABET)

Usher

Though nickelodeons often had attendants to keep order in the entryway, the usher's heyday came in the era of the picture palace. Every picture palace had a corps of uniformed and drilled ushers who performed like a semi-military unit, lining up for inspection before their shifts and marching to their stations. They showed patrons to their seats, answered questions, and handled emergencies, all with the greatest politeness. Today the usher's job is primarily to take tickets.

1 *Lining up for inspection at Radio City Music Hall, c. 1949.*

2 *Employees of the New England Theatres Operating Corporation show off their uniforms, 1929.*

1

2

A highly skilled engineer with extensive technical and practical knowledge of video technology, the video operator is largely responsible for the quality of the images that appear on home television screens. Seated in the control room before a bank of monitors and camera control units, he continually adjusts hundreds of knobs and buttons to balance color values and tones and to achieve a visual uniformity among the multiple cameras. This "image matching" of one camera to another is a continuous process. The associate director cues the video operator whenever a camera change will be made, and the operator then matches the on-deck camera to the one on the air.

The video operator is responsible to the technical director and is required to oversee the maintenance and smooth operation of control room video equipment, cameras, and lenses. (IATSE, IBEW, NABET)

3 In the control room at NBC, 1948.

3

The basic principles of videotape editing are much the same as film editing – to smoothly integrate individual sequences into a unified whole. The technology, of course, is very different. Unlike film editing, videotape editing does not require the physical cutting and splicing of the tape; at its simplest, videotape editing is achieved electronically through the use of machines capable of playback and recording. The editor plays back the unedited tape, selects the precise segment of the tape he wishes to use, and instructs the machine to record (or more appropriately, re-record) that portion of the original tape. Although it is a process of image duplication, steady advancements in the quality of both the videotape and the editing equipment have made the resultant drop in quality practically invisible.

The variety of formats in television programming has created a broad range of applications of the videotape editor's skills. In a narrative format such as a situation comedy, the editor works with the director, producer, or perhaps the associate director to shape the quantities of taped footage into a finished program that fits its time slot to the second. A news program has quite different requirements: The editor usually works closely with the producers and writers to tailor news tape (tapes provided by mobile crews or recorded directly from satellite feeds) to the time period allotted during the broadcast.

The techniques of videotape editing are also finding application in film production. Lured by the relative ease and speed of the process, more and more filmmakers shoot their projects on film, transfer the film to videotape for editing, then cut the original film negative to match the version edited on videotape. (IATSE, IBEW, NABET)

The care and maintenance of a film or television production's costumes and costume accessories are the responsibility of the wardrobe supervisor. During pre-production the supervisor usually attends production meetings with other department heads and prepares a costume breakdown listing scene-by-scene costume and accessory requirements. Working under the costume designer, he or she organizes all costumes, labeling them by actor. The wardrobe supervisor may also go out shopping for smaller items, accessories, or even underwear, as arranged with the designer, and may be involved in fittings with the actors or perform alterations.

The supervisor is responsible for maintaining the costumes, cleaning or repairing them as necessary, and is on the set or location at all times. He or she makes sure that each actor's costumes are ready for each shooting day and helps them dress if they require assistance.

Another major responsibility is continuity: The supervisor takes Polaroids of an actor in costume and keeps detailed notes of the way costumes are worn, and with what accessories since most films are shot out of narrative sequence and the actors' costumes (down to the number of buttons open on a shirt front) must match perfectly when the shots are edited together.

On most productions there is one male and one female wardrobe supervisor who split their duties between male and female cast members. (IATSE, NABET)

A last minute pressing for Nan Wynn on the set of Pardon My Sarong, 1942.

The least visible member of the creative team that determines the character, shape, and quality of the finished production is, perhaps, the writer. Yet the importance of the writer's contribution cannot be underestimated: Bad film and television productions can and have been made from good scripts, but the reverse is very rare. In feature films, the writer is often hired before the director. Writing for television is, by contrast, a continuous process requiring a staff of in-house writers and, when needed, freelance writers may be hired.

The writer ultimately turns the producer's property — novel, play, or story idea — into the final shooting script. He or she works closely with the producer and the director and usually begins by writing a treatment — a detailed narrative account of all the principal situations and sometimes key passages of dialogue. From this the writer develops the finished script, submitting it for approval in stages, and executing any rewrites as requested by the producer or director.

It is common for scripts to be written in collaboration — with the producer, director, or another screenwriter — and, to the legendary chagrin of many writers, it is also common for a completed script to be significantly altered before it reaches the screen. In the best of all

1

situations, a writer will remain on a production, or at least be available for consultations and rewrites, through the end of shooting, but it is just as likely that his or her involvement ends before shooting begins. (WGA)

1 **Dorothy Parker** and **Alan Campbell**, 1936.

2 **Mel Brooks** (at rear) with (left to right) **Ron Clark**, **Barry Levinson**, and **Rudy DeLuca** on the set of *High Anxiety*, 1977.

3 **Ben Hecht** (standing) and **Charles MacArthur**, *Crime Without Passion*, 1934. They also co-directed.

4 **Anita Loos** (seated at left) watches Lewis Stone and Jean Harlow play a scene in *Hold Your Man*, 1933, with Director of Photography Hal Rosson (at camera) and Director Sam Wood (to Loos' right).

3

4

Photo Credits

American Museum of the Moving Image: cover, p. 21 (4), p. 25 (3), p. 49 (3) gift of Dale and Nina Blackwell; p. 6 (1); p. 7 (2); p. 8; p. 11 (2), p. 16 (2), p. 20 (3), p. 41 (1), p. 65 (2), p. 100 (1), (2), p. 112 (1) photo by Muky Munkacsi, gift of Ronald A. and Randy P. Munkacsi; p. 19 (2) courtesy of Concord/New Horizons, gift of Millimeter Magazine; p. 19 (3) courtesy of Paramount Pictures Corporation, The Elephant Man Copyright ©1980 by Brooksfilms Limited, All Rights Reserved, gift of Millimeter Magazine; p. 19 (4), p. 39 (4), p. 87 (4) photo by S. Karin Epstein, p. 90 (1), p. 108 (1) Copyright © by Universal Pictures, a Division of Universal City Studios, Inc. Courtesy of MCA Publishing Rights, a Division of MCA Inc., gift of Millimeter Magazine; p. 22 (1) photo by Barry Wetcher; p. 30 (1), p. 53 (2), p. 62 (1), p. 89 (2) gift of IATSE; p. 43 (6) courtesy of Susan Seidelman; p. 44 (1) courtesy of Columbia Pictures, Nickelodeon ©1976 Columbia Pictures Industries, Inc. All Rights Reserved, gift of Millimeter Magazine; p. 44 (3) courtesy of Inter Planetary Pictures, gift of Millimeter Magazine; p. 46 (1) courtesy of Columbia Pictures, All That Jazz ©1979 Columbia Pictures Industries, Inc. All Rights Reserved, gift of Millimeter Magazine; p. 47 (4) gift of Millimeter Magazine; p. 50 (2), p. 67 (2) photo by Jack Shalitt, gift of Fay R. Shalitt; p. 51 (4); p. 53 (3), p. 69 (2), p. 99 (2) photo by Robert S. Smith; p. 56 (2) gift of Sol Negrin; p. 57 (4) photo by Bruce McBroom, gift of Millimeter Magazine; p. 76 (1) gift of Lester A. Binger; p. 90 (2), p. 96 (1), (2) gift of Christine Jacobsen; p. 98 (1) gift of Millimeter Magazine; p. 105 (2) Copyright © by Universal City Studios, Inc. Courtesy of MCA Publishing Rights, a Division of MCA Inc., gift of Millimeter Magazine; p. 109 (2) gift of Ronald A. and Randy P. Munkacsi; p. 120 (1) gift of William Savoy; p. 120 (2); p. 124 (2) photo by Elliot Marks, from the Twentieth Century Fox release High Anxiety ©1977 Twentieth Century Fox Film Corporation, All Rights Reserved, gift of Millimeter Magazine; p. 125 (3).

The Bettman Archive: p. 18 (1); p. 23 (3); p. 26 (2); p. 43 (7); p. 50 (1); p. 63 (3); p. 81 (3); p. 86 (1); p. 97 (3).

Marc Wanamaker/Bison Archive; p. 7 (3); p. 25 (4); p. 48 (2); p. 70 (1); p. 71 (2); p. 80 (1) photo by Ira Hoke, (2); p. 86 (2).

John Cocchi Collection: p. 104 (1).

Culver Pictures: p. 17 (3); p. 20 (2); p. 24 (2); p. 27 (4); p. 28 (2); p. 33 (3); p. 38 (1), (2); p. 41 (3); p. 42 (1); p. 45 (4); p. 48 (1); p. 50 (3); p. 53 (1); p. 54 (1), (2); p. 56 (1); p. 57 (3) photo by M. Marigold; p. 58 (1); p. 60 (1) photo by Morgan; p. 64 (1); p. 66 (1); p. 77 (2); p. 82 (1); p. 83 (3) photo by Schuyler Crail; p. 88 (1); p. 90 (3); p. 94 (1); p. 101 (3); p. 107 (1), (2); p. 113 (2); p. 125 (4).

The Kobal Collection: p. 11 (1); p. 12; p. 14; p. 20 (1); p. 26 (1); p. 27 (3); p. 28 (1); p. 29 (4); p. 31 (2), (3); p. 32 (1); p. 36 (1), (2); p. 37 (3), (4); p. 42 (3); p. 43 (4), (5); p. 46 (2); p. 60 (2); p. 74; p. 92 (1); p. 93 (2) photo by M. Marigold, (3) photo by Bell; p. 95 (3), (3); p. 99 (3); p. 123.

Richard Koszarski Collection: p. 116, photo by J.C. Milligan.

Library of Congress: p. 59 (2); p. 62 (2); p. 91 (4); p. 110.

Museum of Modern Art/Film Stills Archive: p. 22 (2).

Phototeque: p. 16 (1); p. 29 (3), (5); p. 33 (2); p. 34 (1); p. 35 (2), (3); p. 39 (3); p. 41 (2); p. 42 (2); p. 43 (8); p. 44 (2); p. 45 (5) photo by Brian Hamill; p. 47 (3); p. 67 (3) photo by Schuyler Crail; p. 68 (1); p. 81 (4); p. 82 (2); p. 83 (4), (5); p. 84; p. 87 (3); p. 124 (1).

UPI/Bettman Newsphotos: p. 24 (1); p. 115; p. 121 (3).

Acknowledgments

This book, like the subject it attempts to illuminate, was a collaborative effort from its inception. Rochelle Slovin conceived the original idea and effectively steered the project around a number of obstacles that threatened its progress. Without her support and determination, Behind the Screen could not have been published. Other Museum staff members were also very helpful: Amy Waterman, Eleanor Mish, Richard Koszarski, and Sharon Blume have my sincerest appreciation for their important contributions to this project.

I wish to thank Georgette Hasiotis for her sensitive efforts to reshape an unwieldy manuscript. Stephanie Tevonian met a complex design challenge with great skill. Shari Segel amassed a wealth of photographic material that has contributed enormously to the quality of this book. I would also like to thank many people at photo archives: Marc Wanamaker at Bison Archive, Bob Cosenza at The Kobal Collection, Howard Mandelbaum at Phototeque, Harriet Culver at Culver Pictures, and Rich Wandel at The Bettman Archive.

Finally, I owe great appreciation to all the people at the unions and working within the film and television industries who generously provided guidance and information and took the time to answer questions about their work. They and their colleagues were the inspiration for this book, and it would have been impossible without their help. I would especially like to thank Henry Alford, Richard Allen, Stewart Allen, Barbara Austin, George Barimo, Ed Beyer, Susan Bode, Conrad Brink, Herman Buchman, John Caglione, William Christians, John Corso, Marko Costanzo, Cliff Cudney, Richard Dean, Joe DiBesey, Lee Dichter, Karin Epstein, Peter Fallon, Ed Fanning, Harriet Fidlow, John Finnerty, Wayne Fitzgerald, Romaine Green, Shelly Houis, Vito Ilardi, Larry Kaplan, John Kasarda, Jeffrey Kurland, Walt Levinsky, Barbara Matera, Jim Mazzola, Barbara Minsky, Andrew Mondshein, Sol Negrin, Milton Olshin, Eddie Quinn, Tom Razzano, Joseph Reidy, Samuel Robert, Barbara Robinson, Dan Sable, Mark Schubin, Eoin Sprott, Jo Umans, Richard Ventre, and Connie Wexler. —David Draigh.

Index to Job Titles

Actor, 6
ADR Editor, 9
Advisor/Consultant, 10
Agent, 10
Animal Trainer, 11
Animator, 12
Apprentice Film Editor, 55
Apprentice Sound Editor, 114
Art Director, 13
Assistant Art Director, 13
Assistant Costume Designer, 37
Assistant Director, 14
Assistant Film Editor, 55
Assistant Music Editor, 73
Assistant Property Master, 94
Assistant Sound Editor, 114
Associate Director, 15
Atmospheric Effects Specialist, 16
Audio Operator, 17
Best Boy, 18
Boom Operator, 18
Cable Puller, 20
Camera Operator (Film), 21
Camera Operator (TV), 23
Carpenter, 25
Casting Director, 26
Caterer, 27
Choreographer, 28
Clapper/Loader, 101
Color Timer, 30

Composer, 31
Concessions Worker, 32
Conductor, 33
Construction Coordinator, 34
Construction Foreman, 34
Costume Designer, 35
Costume Maker, 38
Crane Operator/Grip, 39
DGA Trainee, 40
Dialogue Coach, 40
Director, 41
Director of Photography, 45
Dolly Grip, 48
Draftsperson, 48
Driver, 49
Electrician, 50
Executive Producer, 51
Extra, 51
Extra Casting Director, 52
Film Editor, 53
First Assistant Camera Operator, 56
Focus Puller, 56
Foley Artist, 57
Gaffer, 58
Generator Operator, 58
Grip, 59
Hairstylist, 60
Key Grip, 61
Lab Technician, 62
Lighting Director, 63

Location Manager, 64
Location Scout, 65
Louma Crane Operator, 65
Makeup Artist, 66
Matte Artist, 68
Mechanical Effects Designer, 69
Miniature Designer, 70
Model Maker, 72
Music Arranger, 72
Music Editor, 73
Music Recording Supervisor, 74
Musician, 74
Musician Contractor, 75
Negative Cutter, 76
Nurse/Paramedic, 77
Optical Printer Operator, 78
Playback Operator, 79
Post-Production Supervisor, 79
Process Projectionist, 80
Producer, 81
Production Assistant, 84
Production Associate, 85
Production Auditor/Accountant, 85
Production Designer, 86
Production Illustrator, 110
Production Manager, 88
Production Office Coordinator, 89
Production Sound Mixer, 90
Projectionist, 91
Property Maker, 92

Property Master, 93
Prosthetic Makeup Artist, 95
Pyrotechnic Specialist, 96
Recording Engineer, 97
Re-Recording Mixer, 97
Scenic Artist, 99
Script Supervisor, 100
Second Assistant Camera Operator, 101
Second Assistant Director, 102
Second Unit Director, 102
Set Decorator, 103
Set Dresser, 103
Sideline Musician, 104
Sketch Artist, 110
Special Effects Coordinator, 105
Stagehand, 106
Stage Manager, 106
Stand-In, 107
Steadicam Operator, 108
Still Photographer, 109
Storyboard Artist, 110
Stunt Coordinator, 111
Stunt Person, 112
Supervising Sound Editor, 113
Technical Director, 115
Theater Owner/Manager, 116
Theme Song Composer, 116
Title Designer, 117
Transportation Coordinator, 118
Treatment Writer, 118

Unit Publicist, 119
Usher, 120
Video Operator, 121
Videotape Editor, 122
Visual Effects Supervisor, 105
Wardrobe Supervisor, 123
Writer, 124